An Introduction to
Theoretical Geomorphology

TITLES OF RELATED INTEREST

Aeolian geomorphology
W. G. Nickling (ed.)

Catastrophic flooding
L. Mayer & D. Nash (eds)

Discovering landscape in England and Wales
A. S. Goudie & R. Gardner

Environmental change and tropical geomorphology
I. Douglas & T. Spencer (eds)

Experiments in physical sedimentology
J. R. L. Allen

The face of the Earth
G. H. Dury

Geomorphology: pure and applied
M. G. Hart

Geomorphology in arid regions
D. O. Doehring (ed.)

Geomorphology and engineering
D. R. Coates (ed.)

Geomorphological field manual
V. Gardiner & R. Dackombe

Geomorphological hazards in Los Angeles
R. U. Cooke

Geomorphological techniques
A. S. Goudie (ed.)

Glacial geomorphology
D. R. Coates (ed.)

Hillslope processes
A. D. Abrahams et al. (eds)

The history of geomorphology
K. J. Tinkler et al. (eds)

Image interpretation in geology
S. Drury

Karst geomorphology and hydrology
D. C. Ford & P. W. Williams

Mathematics in geology
J. Ferguson

Models in geomorphology
M. J. Woldenberg et al. (eds)

Planetary landscapes
R. Greeley

A practical approach to sedimentology
R. C. Lindholm et al. (eds)

Principles of physical sedimentology
J. R. L. Allen

Rock glaciers
J. Giardino *et al.* (eds)

Rocks and landforms
J. Gerrard

Sedimentology: process and product
M. R. Leeder

Tectonic geomorphology
M. Morisawa & J. T. Hack (eds)

An Introduction to Theoretical Geomorphology

COLIN E. THORN

Department of Geography, University of Illinois

Boston
UNWIN HYMAN
London   Sydney   Wellington

Allen & Unwin, Inc.,
8 Winchester Place, Winchester, Mass. 01890, USA

Published by the Academic Division of
Unwin Hyman Ltd
15/17 Broadwick Street, London W1V 1FP, UK

Allen & Unwin (Australia) Ltd,
8 Napier Street, North Sydney, NSW 2060, Australia

Allen & Unwin (New Zealand) Ltd in association with the
Port Nicholson Press Ltd,
60 Cambridge Terrace, Wellington, New Zealand

First published in 1988

Library of Congress Cataloging in Publication Data

Thorn, Colin E.
An introduction to theoretical geomorphology/Colin E. Thorn.
  p.   cm.
Bibliography: p.
Includes index.
ISBN 0-04-551117-9 (alk. paper).
ISBN 0-04-551118-7 (pbk.: alk. paper)
1. Geomorphology.   I. Title.
GB401.5.T47 1988
551.4 – dc19   88-6236 CIP

British Library Cataloguing in Publication Data

Thorn, Colin E.
An introduction to theoretical geomorphology.
1. Geomorphology
I. Title
551.4

ISBN 0-04-551117-9
ISBN 0-04-551118-7 Pbk

Typeset in 10 on 12 point Times by Computape (Pickering) Ltd, N. Yorks
and printed in Great Britain by Biddles of Guildford

To my mother and late father:
for many sacrifices and much encouragement
To Carole:
for endless patience and support
To Stephen and Jeffrey:
for insightful indifference and joie de vivre

Acknowledgments

Geomorphology has provided me with a great deal of enjoyment and stimulation. In very large part the opportunity to pursue an academic life was given to me by a number of gifted and attentive mentors. My run of good fortune in education started at Maidstone Grammar School for Boys: I can think of few places that could match the quality and quantity of opportunities it provided students such as myself.

At Nottingham University, Cuchlaine A. M. King, and John C. Doornkamp initiated and steered my early interest in geomorphology. Thereafter, Bruce G. Thom at the McGill Sub-Arctic Research Laboratory introduced me to the delights of frigid fieldwork. This was followed by a year-long course in glaciology under the late Fritz Müller. While glaciology remained a tangential interest, Professor Müller's attitude towards science has always served me as an inspiration.

As a student, my studies culminated at the Institute of Arctic and Alpine Research, University of Colorado at Boulder. The vibrancy of those days is difficult to convey, but will always endure in my mind. John T. Andrews, Roger G. Barry, T. Nelson Caine, Jack D. Ives, and many graduate colleagues created an intellectual environment as dazzling as, and much more penetrating than, Boulder's campus and streets. I think that I, and most other mountain geomorphologists, owe Nel Caine a considerable debt for his intellectual leadership. In my own case, I also owe Jack Ives an enormous debt of gratitude for support well beyond the call of duty, opportunities, an overabundance of enthusiasm, and friendship.

Whatever good falls between these covers is, in very large part, because I have had the good fortune to be a student of these scholars. Whatever ills and errors remain I have certainly created myself!

I would also like to thank the students at the University of Maryland, College Park, and the University of Illinois at Urbana-Champaign, who have suffered oral versions of this book in a graduate seminar. Deborah S. Loewenherz emerged from such an experience sufficiently unscathed to render great service in the preparation of Chapter 13, which contribution is gratefully acknowledged here.

Nel Caine (University of Colorado at Boulder), Ronald U. Cooke (University College, London), and William L. Graf (Arizona State University) made many helpful comments on draft chapters. Their efforts were much appreciated. Jim Petch (University of Salford) also provided helpful remarks about systems modeling.

Barbara B. Bonnell typed the manuscript with accuracy, speed, and good grace that defies the hubbub that passes for our departmental office. Without her efforts, this book may never have emerged!

It would be an injustice not to comment on the library of the University of Illinois at Urbana-Champaign. It is truly one of the great research tools in the United States; unfortunately, its good health seems to be perpetually threatened, but to date it has survived. Having suffered libraries elsewhere, it has been a true delight to wallow in the seemingly bottomless font of human knowledge that rests on the shelves in Urbana-Champaign.

A few figures were drawn or redrawn specifically for this text by James Bier (University of Illinois at Urbana-Champaign).

We are grateful to the following individuals and organisations who have kindly given permission for the reproduction of copyright material (figure numbers in parentheses):

R. Abler (2.1, 2.2, 2.3); Edward Arnold Publishing and D. Harvey (2.4a); Paul Chapman Publishing and R. Haines-Young (2.4b); Edward Arnold Publishing, D. Harvey, P. Caws, reproduced by permission of Wadsworth, Inc (4.1); Geological Society of America (Table 4.1); A. N. Strahler (5.1); *American Journal of Science* and S. A. Schumm (6.1, Tables 6.1 and 6.2); Edward Arnold Publishing and K. J. Gregory (6.2); Longman and P. J. Williams (6.3); *American Journal of Science* (6.4); R. J. Chorley and B. A. Kennedy (6.5, 12.7); Geological Society of America and W. R. Womack (6.6); A. Getis, reproduced by permission of Cambridge University Press; Methuen & Co, M. F. Dacey and E. H. Thomas (7.2); Association of American Geographers and R. A. G. Savigear (8.1); *The Journal of Geology*, by permission of the University of Chicago Press, and R. J. Chorley (8.2); Canadian Association of Geographers and E. D. Ongley (8.3); Geological Society of America and A. N. Strahler (8.4); Royal Geographical Society (10.1); Association of American Geographers and G. H. Dury (10.2, 10.3, 10.6, 10.8); Longman and A. Young (10.4, 10.10); Macmillan (10.5); Figure 10.7 reproduced by permission of John Wiley & Sons, Ltd; Methuen & Co and J. B. Thornes (10.9); Association of American Geographers & W. H. Terjung (12.1); The Institute of British Geographers and C. T. Agnew (Table 12.1); Figure 12.3 by permission of Barnes & Noble Books, Basil Blackwell Publisher and C. A. M. King; *The Journal of Geology*, by permission of the University of Chicago Press, and M. A. Melton (12.4, 12.5, Tables 12.2, 12.3 and 12.4); Methuen & Co, R. J. Chorley and B. A. Kennedy (12.6); V. H. Winston & Sons, Inc and A. N. Strahler (12.8, 12.9, Tables 12.5 and 12.6); M. A. Carson, reproduced by permission of Cambridge University Press (13.1); Université de Liège and F. Ahnert (13.2, 13.3, Tables 13.1 and 13.2); The University of Maryland College Park and F. Ahnert (13.4); *Zeitschrift für Geomorphologie* and M. J. Kirby (13.5); M. J. Kirby, reproduced by permission of John Wiley & Sons (13.6); The Institute of British Geographers and M. J. Kirby (13.7); The Geographical Association and J. B. Thornes (14.1, 14.2).

Contents

Tables

An Introduction to
Theoretical Geomorphology

1 Introduction

This book stems from the belief that undergraduate, and many graduate, students in geomorphology are taught far too little about the theoretical foundation of the discipline. This is a tragic oversight because all disciplines exist solely by virtue of a body of theory; furthermore, all substantive research must, of necessity, be directed at improving this body of theory. While the centrality of theory is a characteristic of all science, it appears that it is fieldwork that has achieved sacrosanct status in geomorphology. The supreme standing of fieldwork would seem at first sight to be supported by such comments as that cited by Ager (1984, p. 42): "My old professor – the great H. H. Read – once said, 'The best geologist is, other things being equal, he who has seen the most rocks'." The key issue in Read's comment is obviously "other things being equal"; what are these other things? For they will be just as applicable to geomorphology as to geology at large. One is certainly personal ability, but another critical component is the individual's knowledge of relevant theory. Pursuit of theoretical knowledge is not generally a favored activity in geomorphology, a situation epitomized by Chorley's (1978, p. 1) one-liner: "Whenever anyone mentions theory to a geomorphologist, he instinctively reaches for his soil auger."

Geomorphology is not a discipline without theory; indeed, such a statement is quite literally nonsense, whether applied to geomorphology or any other discipline. However, geomorphology does appear to be a discipline in which many practitioners seem unaware of theory, consciously consider it unnecessary, or view it as a pretentious embellishment. The old saw that "geography is learnt through the soles of your feet" is clearly also a widely accepted view of the way to approach geomorphology. Chorley (1978, p. 1) once again provided a pithy summary of reality, commenting that "the only true prisoners of theory are those who are unaware of it."

Theoretical geomorphology is presently growing at a rapid pace, but in an inefficient fashion. The problem, and there is a problem, is the slowness with which theory, both old and new, is disseminated through the discipline. A small group of eminent scholars are making significant contributions to theoretical geomorphology, but their published papers to a considerable extent represent "personal communications" to each other because theoretical work is considered tangential to the mainstream of geomorphology by a high proportion of geomorphologists.

The intent here is not to decry the importance or significance of fieldwork – this would be inappropriate because knowledge of the Earth's surface is still woefully incomplete. Rather, the purpose is to suggest that fieldwork, and hence geomorphology itself, is currently being sold short because so much of it is conducted without adequate and prior consideration of relevant theory. Furthermore, the sequence of the two is critical, theory must precede fieldwork.

The specific audience to whom this volume is addressed is undergraduate students and beginning graduate students. Most of these geomorphologists in the U.S.A. are rarely asked to consider theoretical issues in their coursework. As a result, many of them confront theoretical issues for the first time when trying to develop a research topic or, not uncommonly, as a "hindcasting" procedure when trying to write a thesis or dissertation having already collected the data without much regard for theoretical issues beforehand. Therefore, this text is not an attempt to make a contribution to geomorphic theory *per se*, but rather seeks to disseminate existing geomorphic theory in a fashion that is comprehensible and useful to the undergraduate and beginning graduate students who will form the next generation of research geomorphologists. As a result, this book is long on synthesis and short on originality.

Content

Any attempt by a single individual to create a comprehensive review of geomorphic theory is going to be constrained and flawed by the limitations of personal experience and abilities, and this volume is certainly no exception. Therefore, at least some of these limitations should be identified. Most important of all is the fact that this is almost exclusively an examination of geomorphic theory produced in the English-speaking world. This is of the utmost importance because geomorphology is a discipline that still exhibits very distinctive national schools.

Today most English-speaking geomorphologists (including the writer) have been, or are being, trained as process geomorphologists. As a result, this book is founded on the body of theory that prevails in process geomorphology. There are some excursions into other domains, but they fall far short of fairly representing other approaches to geomorphology.

As a matter of choice, this book is devoted to theory and methodology, and not to philosophy and techniques. These are terms that are often confused in everyday conversation but should be sharply distinguished. All are discussed in Chapter 2, but the above distinction may be restated for the time being as: this book is about concepts and internal logic, and not about what geomorphologists should study or about the precise procedures they should use to generate and evaluate data.

At a much less important level it is worth identifying the content of the book from what might be called a stylistic point of view. Many "classic" papers are used, not only as sources of ideas but also for many of the examples cited. As a result, the reader will find very little that reflects or incorporates contemporary techniques. This strategy serves two purposes: first, it reinforces the contributions made by a few truly eminent geomorphologists who have done the most to shape post-World War II geomorphology; second, it focuses attention upon the ideas themselves as opposed to complicating the task by also presenting complex techniques.

A final aspect of the content of the book is that it is overwhelmingly qualitative, rather than quantitative. This partly reflects the writer's own abilities, but also the fairly low level of mathematical expertise that the intended audience exhibits. Mathematics is often called the language of science, and so it is true that theory is best expressed mathematically. However, mathematics is only a language and, therefore, theory is not itself intrinsically mathematical. It is hoped that what is lost in precision by abandoning mathematics will be offset by increasing the size of the audience.

Organization

The book is subdivided into two unequal parts. The first part is devoted to a series of individual topics or themes that are central to contemporary geomorphic theory. It is unrealistic to isolate them from each other in this fashion, but heuristically unavoidable.

Part One begins (Chs. 2–4) with an attempt to provide a context within which to consider geomorphology. This is pursued initially by a brief consideration of science in general. Following this is an examination of the various forms in which geomorphology appears as an academic discipline. These preliminary steps are concluded with a review of terminology and its links with theory.

Chapters 5–8 embrace most of the important themes in space, time, and scale that confront geomorphologists. Clearly, spatial–temporal interaction is central to geomorphology, but while recognizing the importance of this interaction, separate treatments of individual concepts provide much clearer, if simplistic, introductions. Chapter 9 is a summary of Part One in which a number of the individual threads are drawn together, although the task is far too difficult to be treated in an entirely satisfactory fashion.

Part Two of the text (Chs. 10–14) is devoted to past and present models of landscape evolution. These are treated individually (Chs. 10–13), and range from W. M. Davis's (1899) geographical cycle to M. J. Kirkby's

(1986) mathematical model. Part Two of the text also ends with an attempt (Ch. 14) to integrate some of the issues treated separately in the section to that point; again the complexities at hand are so great that they cannot be resolved in anything like a comprehensive fashion.

A disciplinary context

Given the stated purpose of this text, it seems appropriate to suggest a context within which the student may view it and the themes that complement it. There are many possible intellectual goals in the study of landforms; accordingly there are many disciplinary philosophies. These are matters of personal preference and cannot be labeled right and wrong. However, there are other matters that seem worthy of promotion because they are applicable whatever type of geomorphologist the student chooses to become.

In the first instance most students fail to develop much appreciation of the history of the discipline. There is no particular need to develop such knowledge to an extreme degree, but on the other hand it is very hard to comprehend where the discipline is headed if one has no idea of from where it has come. A number of excellent sources are cited in the chapters ahead, but a reading of something as brief as Tinkler's (1985) text should be mandatory.

This text itself explores only one portion of a much broader and deeper scene and that is the entire nature of science itself. Most geomorphologists like to think of themselves as scientists, but few appreciate the fragility of the calling, or care to consider carefully the rules of the game they claim to be playing. Haines-Young & Petch (1986) is an excellent review of many of these profound issues: a careful reading of it should cause sufficient alarm to induce the necessary sense of trepidation with which to offset the temerity that generally comes much more readily.

It is important to re-emphasize that the intention in this text is not to demean fieldwork; in fact, exactly the opposite is true – the intent is to show that fieldwork can be much more valuable when it is guided by, and rests upon, a previously established theoretical foundation. The importance or role of fieldwork pursued without a theoretical complement would seem to be largely recreational and therapeutic and consequently meritorious, but irrelevant professionally.

Finally, every student should be urged to pursue the language of mathematics urgently – once he or she has accepted it as a language that may be "spoken" well or badly, and has recognized it as a medium that usually insures logic and rigor, but guarantees neither. Most techniques, mathematical or otherwise, need to be characterized in a similar vein because they have become the tail that often wags the dog. A sophisticated technique is a powerful tool in the hands of an artisan who has chosen it for its relevance to the task at hand; a "sophisticated" technique applied inappropriately is neither sophisticated nor powerful – it is merely seductive. This set of ideas has been expressed much more eloquently by Winkelmolen (1982, p. 264):

> Mathematics and statistics are powerful tools for geologists, but we should not use them to put up a smoke-screen of quasi-exactness. A computer is not a washing machine in which our data can be purified. We should rather use the time that can be saved by using it to reflect on our basic suppositions.

The spirit of this text is to commend geomorphology to students as a thinking-before-digging (or looking) discipline. Theory is an integral part of being human, whether applied to daily life, geomorphology, or any other human endeavor. Given that theory is as essential intellectually as air, water, and food are physically, to ignore it, or to label it superfluous and/or pretentious, is the height of foolishness; primarily because it permits theory to control thinking unwittingly. On the other hand, to pursue it renders theory malleable and, even more importantly, recognizable for the fragile and transient man-made structure that it is. With a little good fortune, what follows may serve as a starting point for some of those who find coasts and mountains, and even east-central Illinois, as fascinating as the writer does and wish to try to understand their development.

Part one

2 Science – the reality

On thinking

The human brain is so highly developed that its good health depends upon a constant stream of experiences for it to organize and order. An experience is our perception of an event, and probably the most fundamental steps in ordering experiences are to locate them in time and space. As spatial and temporal issues are central to all disciplines, and not just to geomorphology, they are worth looking at in very general terms.

Figure 2.1 provides a schematic summary of variations in impact derived from the timing and location of experiences. Temporally, what we are experiencing at any moment is both very intense and narrow (technically, only one thing at a time); in addition, the past is much more influential than the future. Spatially, the obvious characteristic is the influence or importance of rare experiences, and although Figure 2.1d is schematically shown as circular it is really highly channelized.

Following Abler *et al.*'s (1971) development of the general thought process, it is also possible to represent schematically the manner in which we order experiences (Fig. 2.2). Constructs always have empirical content and are used to define or categorize experiences. Close to the P-plane (P for protocol, perception or primary; Margenau 1961, p. 5) constructs are very specific. but with increasing distance from the P-plane there is increasing generalization. In contrast to constructs, concepts lack empirical content. The role of concepts is to permit us to manipulate constructs. Concepts are hierarchical or ordered, and among the most important are the "mega-concepts" of number and relationship. In fact, number and relationship (the latter clearly embracing time and space) are so important that they form the basis of the two formal sciences of logic and mathematics. Logic and mathematics are distinguished from all other sciences by their lack of specific subject matter.

Creation of linkages between constructs and concepts, and particularly between concepts and concepts, affords a dazzling array of possibilities. The number is too great for all links to be made automatically and we have a very strong tendency to be consciously selective. This means that we have needs to satisfy and, because these may very well vary, we often require

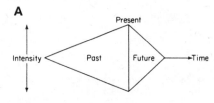

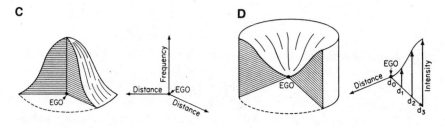

Intensity of experience in the temporal context.   *Variety of experience in the temporal context.*

Frequency of experience in the spatial context.   *Intensity and variety of experience in the spatial context.*

Figure 2.1   Human experiences in temporal and spatial contexts. (From Abler *et al.* (1971).)

multiple views of any construct or concept. In addition, we have some preconceived notions of what constitute meaningful relations. In short, we have ordering systems.

There are several ordering systems available to us; for example, theology, esthetic, common sense, and science (Abler *et al.* 1971). As if this were not enough, science itself is far from being a monolithic concept (e.g., Harvey 1969). Furthermore, even within the confines of a small discipline such as geomorphology, we find great diversity of objectives and theoretical underpinning (e.g., Chorley 1978).

Science

Above all else, it is critical to realize that, whatever "science" is, it is the creation of man. This needs to be emphasized because there is a common tendency to endow it with a life force of its own, rather as if it came from another planet, already refined, prepackaged, monolithic, and entirely unassailable. Quite the contrary is true; what is considered *bona fide* science varies with the times, differs from place to place, and is the topic of sufficient debate that the history of science is a lively research field in its own right.

Science is a powerful ordering system, very much an artifact of Western

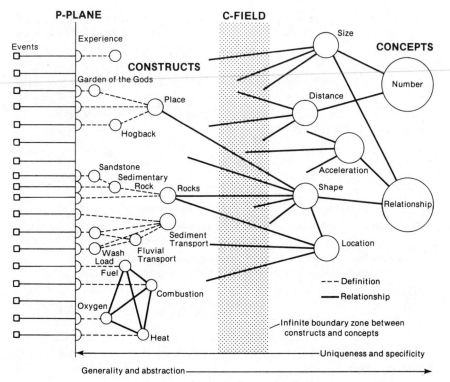

Figure 2.2 Human organization of experience. (From Abler *et al.* (1971).)

culture (Abler *et al.* 1971, p. 24), and commonly expressed in its most refined form in the language of mathematics. Any attempt to characterize science and its constituent parts requires use of some terms that are commonly used rather loosely in colloquial speech, and even by many scientists. This is unfortunate as it robs us of some important distinctions that are well worth recognition.

Philosophy

The philosophy of a discipline is the set of beliefs held about that discipline. Consequently, a disciplinary philosophy is the ultimate source of what workers in the discipline consider to be worthwhile objectives. Harvey (1969) pointed out that beliefs may not be invalidated by arguments based on their logic, or lack thereof. If one geomorphologist believes that the object of geomorphology is to reconstruct Tertiary landscapes, another to model Pleistocene ice sheets, and a third to manage spoil-tip reclamation, it is reasonable to expect very different geomorphologies to emerge. Nevertheless, it is not possible to argue that any one of these beliefs or objectives is inherently illogical.

If one chooses to reject research at this level, it must be on the grounds that it is not relevant to one's own objectives. Obviously, the message here is to develop clear ideas about one's own objectives and equally clear ones about the objectives of any author. Acceptance or rejection of the work of others on the grounds of relevance or irrelevance is an external test, and the first step in a two-step procedure. The second step is an internal test; namely, to see if the objective is treated in an internally logical fashion. This second step takes us out of the realm of disciplinary philosophy and into methodology.

Methodology

The internal logic, or methodology, of a discipline stems from theory. In the natural sciences, physics stands as the pre-eminent discipline in theoretical refinement and invariably it has been the norm against which other disciplines have been measured. Inasmuch as geomorphology may be viewed as the chemical and physical behavior of crustal materials at the Earth's surface, physics is clearly an appropriate yardstick for geomorphologists. However, there is sufficient conflict over the way in which theory and methodology should be developed to urge anyone to regard the norm of physics as a lodestone rather than a touchstone. Many of the issues and conflicts are elegantly addressed by Harvey (1969) and Haines-Young & Petch (1986).

Theory

Harvey (1969, p. 88) stated: "A theory is, thus, a system of statements". Formally, a theory is entirely abstract (i.e., without empirical content) and is derived from "primitive terms" (nondefinable items) that are used, in turn, to derive axioms and then theorems. Use of primitive terms, axioms, and theorems must be in accordance with specified rules of operation (invariably those of deduction). Synthesis of all these elements produces a calculus (Harvey 1969, p. 88). So far, this entire package is abstract and it is only useful in empirical sciences if it can be successfully related to empirical phenomena.

The critical link that relates the abstract to the empirical is called a text. A text must fulfill two requirements; it must interpret abstract theory in terms of empirical observations, and it must define the domain (area or sector) over which the theory is considered applicable. The presence of a calculus, plus a text, are necessary and sufficient conditions to justify calling the theory a scientific theory.

The "statements" in Harvey's definition of a theory are most commonly referred to as laws. A sequence may be envisaged that runs: factual statement → generalization → empirical law → general or scientific law

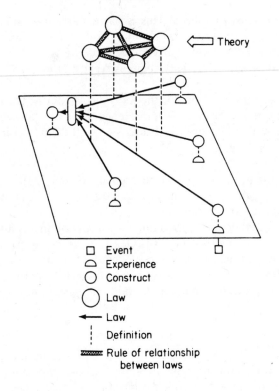

Theoretical structure in the C-field.

Figure 2.3 Relationships between fact, law, and theory. (From Abler *et al.* (1971).)

(Harvey 1969, p. 31). Laws are such a fundamental concept in science that it would be comforting if they were cast in concrete, but they are not.

Conceptually, a law is supposed to be a universal statement devoid of constraints. Harvey (1969, pp. 100–6) elucidated the philosophical problems stemming from this requirement. The majority of scientists function by accepting what is called "methodological universality", that is, if they decide that a statement is reasonably regarded as being law-like, they treat it as one.

In addition to the notion that a law must be universally true, or at the very least widely applicable, there is another basic requirement. Individual statements are not usually regarded as laws, whatever their seeming veracity, unless they can be fitted into a web of laws, that is, into a body of theory. Relationships between facts, laws, and theory are shown schematically in Figure 2.3.

Any high-minded hopes we might harbor for an iron-clad definition of a scientific law must fall short. This is particularly true outside the formal sciences, and certainly overwhelmingly true if one is searching for geomorphological laws. Although various kinds of "law" will emerge in the pages ahead, it is as well to see them as Ackoff (1962, p. 21) did:

> The less general a statement the more *fact-like* it is: the more general a statement, the more *law-like* it is. Hence, facts and laws represent ranges along the scale of generality. There is no well-defined point of separation between these ranges.

Or as Goodman (1967, p. 99) phrased it: "Indeed, the very distinction between fact and theory is one of relative scope; a fact amounts to a small theory, a theory to a broad fact."

Whatever intellectual shortcomings are apparent in any attempt to define laws, and hence theory itself, it is clearly possible to create satisfactory, functional definitions. If theory provides the foundation of methodology, techniques are the tools that, when properly applied, permit us to scrutinize existing theory and to expand and modify it as necessary.

Technique

As Chorley (1972, p. 9) observed " . . . , techniques are merely the artifacts of scholarship". Accordingly, by far the most important measure of a technique is the extent to which it permits us to advance toward a stated goal. As has already been suggested, a geomorphologist may have a variety of goals or objectives and, therefore, techniques. However, even a single goal may be approached by a variety of techniques.

Fieldwork is probably the oldest and most widely used approach in geomorphology and it embraces a number of individual techniques. Observation of form alone has a long tradition, ranging from a simple verbal description through detailed morphological mapping (e.g., Savigear 1965). Spatial analysis (location with respect to both horizontal and vertical position) in a very simple form is also an old field technique. One has only to think of the inferences drawn from the relative position of two landscape components, e.g., a free face and talus. A third field technique is examination of the character or internal attributes of sedimentary deposits (e.g., King 1966). Detailed analysis is often undertaken in a laboratory, but an enormous amount of information can be generated in the field alone. Till fabric analysis (Andrews 1971) and soil descriptions (Birkeland 1984) constitute two areas where good fieldwork may stand alone and, even when supported by laboratory analysis, can never be supplanted by it. All of these techniques are static and may be complemented by dynamic field observations.

Process geomorphology, as usually practiced in the field, is measurement of process rates and their variation in time and space. This may seem self-evident, but it implies that the physics and chemistry of individual processes are rarely studied outside the laboratory. Thus, a process in the field sense is often only fuzzily understood or may, in fact, not be a single mechanism. Conversely, laboratory research that pinpoints mechanisms in detail is often conducted at a scale that makes linkage to field studies difficult. While field process studies are heavily biased toward sediment transport and depositional mechanisms, it is now possible to measure bedrock erosion directly in the field (e.g., High & Hanna 1970), although periods over which this is done may not be very appropriate.

Church (1984, p. 563) defined a scientific experiment as "... an operation designed to discover some principle or natural effect, or to establish or controvert it once discovered". He went on to observe both how loosely the term has been used in geomorphology and how little truly experimental geomorphology has been conducted. Church emphasized establishment of experimental control, either by direct human interference in the landscape or by statistical methods (see also Anhert 1980, Haines-Young & Petch 1986). Slaymaker et al. (1980) have offered a more liberal interpretation of the word "experimental", thereby bringing more geomorphic research into the fold.

Perhaps the most common instance of experimental geomorphology in which direct human interference takes place is in soil loss studies incorporating plots and artificial rain (e.g., various papers in Bryan & Yair 1982). Use of statistical means to establish experimental control has been fairly common in hydrologic studies of drainage basins. Yet another approach to these issues has been use of scale models. Mosley & Zimpfer (1978) discussed a variety of laboratory hardware models which range from the large (e.g., Parker 1977) to the very small (e.g., Glen 1952). In nearly all instances, it is impossible to avoid distorting some aspects of scale relationships between geometric, kinematic, and dynamic dimensions of the model when compared to the prototype (King 1966, p. 188). However, some aspects of dimensional scaling have long been studied.

While laboratory hardware models are designed to mimic some segment of reality, this is only one aspect of contemporary geomorphic research conducted in laboratories. Detailed analysis of characteristics that may be measured only roughly in the field are undertaken with great frequency in the laboratory, grain-size distribution being an obvious illustration. Yet another sphere of laboratory research is direct investigation of physical and chemical properties whose changes underlay more broadly defined field processes; examples include triaxial shear stress and clay mineralogy.

At the opposite end of the spectrum are the multitude of aerial photographic and remote-sensing techniques. Much of this work overlaps older techniques, but addition of sensors that record conditions not detectable by human senses, great capacity for repetition, and ability to digitize information have moved research into realms only dreamt of previously. Much of this progress rests squarely on availability of high-speed computers and advanced quantitative techniques.

Quantitative techniques merit independent recognition. Not infrequently, quantitative geomorphology is called theoretical, but this is a misnomer. It is more appropriate to regard theoretical as a synonym for good rather than for quantitative. All good geomorphology, regardless of technique used, must address theory; conversely, research that does not address theory (either directly or indirectly) must, by definition, be bad. This perspective will be amplified shortly.

Quantitative techniques may be conveniently subdivided into two basic categories, stochastic and deterministic, or statistical and analytical (Mather 1979). In practice, statistical approaches in geomorphology have been dominated by correlation and regression techniques. As Mather (1979, p. 476) indicated, the statistical approach is a "black-box" one geared to prediction. Its primary weakness is that clarification of relationships between processes and responses (landforms) are masked by a complex of interactions. Even from an elementary perspective it is easy to see how difficult it is to isolate anything in the natural world, and to this must be added many problems associated with scale dependence.

Deterministic or analytical techniques are dependent on specification of relationships derived from existing theory (most commonly those of chemistry or physics), followed by deduction of consequences that must result from these relationships. While deterministic approaches are innately appealing, because they offer the prospect of fundamental understanding of cause and effect, they remain fragile. This is due to our ignorance of many important geomorphic relationships, our use of questionable assumptions, and our frequent incorporation of stochastically derived data or steps in deterministic work.

Clearly not every technique may be neatly pigeonholed and most geomorphologists who acquire a high reputation are eclectic in their choice of techniques. However, techniques themselves are frequently so sophisticated that a sizable portion of professional effort and time is channeled into mastery of technique alone. As a result, geomorphologists are often isolated from each other as much by their choice of technique as by their choice of subject matter or objectives. This is unfortunate and unnecessary because techniques may be subjected to a unifying measure; namely, their ability or failure to meet standards of acceptable scientific methodology. Given a common objective and a common measure of acceptability, technical diversity is a strength rather than a weakness. The next step along the road is to examine this measure of acceptability, namely the scientific method.

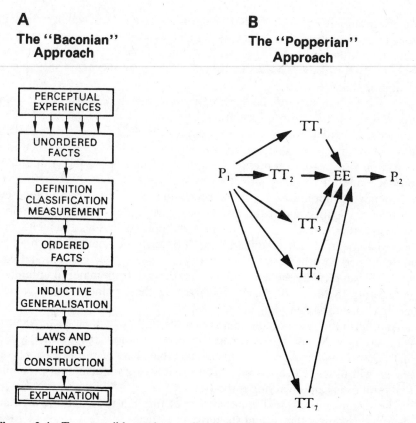

A

The "Baconian" Approach

B

The "Popperian" Approach

Figure 2.4 Two possible pathways for "scientific" reasoning. (Part (a) from Harvey (1969); part (b) from Haines-Young & Petch (1986).)

Scientific Method

Certainly, virtually every geomorphologist aspires to be "scientific", even if he or she has only a very fuzzy notion of what this might be. In many ways, the term has devolved into merely a synonym for logical, thorough, or praiseworthy. As with many other entrenched terms that have achieved sacrosanct status, "scientific" has some rather specific connotations that have been lost in common usage, nor indeed is the term nearly as clear-cut or well-defined as most people assume.

Harvey (1969, p. 31) pointed out that science as an activity contains several disparate elements. First, there is the actual process of discovery, a mental activity. Second, there are certain procedures that are required to generate support for one's conclusions if they are to be accepted by the scientific community at large; this is really a social process. Third, the scientist is required to present his or her findings in an ordered format that meets commonly held standards; again this is really a social activity, but it is specifically a written task. Much of the conflict that surrounds attempts to define scientific method precisely has stemmed from complexity generated by attempting to embrace so many different kinds of activity in a single definition.

Two basic outlines of the supposed route that a scientist takes from initiation to completion of scientific research appear in Figure 2.4. The so-called Baconian route has two important implicit concepts. These are that "facts" may exist independently of any theory, and that by a process of induction (i.e., generalization from a body of unordered facts) one may produce laws, often called inductive laws.

Among the more obvious problems with induction are questions focused on how to identify and define a fact in the supposed absence of theory. Similarly, there must be some preconceived notions that lead to identification of a law from collected facts and, in order to be scientific, such a law must be tied to other laws (i.e., theory). It is a classic chicken-and-egg situation; supposedly the data and law are derived from material untainted by existing theory, but clearly it is existing theory that suggests which material is worth examining at the outset.

Another basic flaw in induction is that the findings can never be viewed with certainty. No matter how many times observations, experiments, and analyses are undertaken, it is always possible that on the next occasion results will differ. In short, there is no underlying fundamental principle that establishes an unvarying response.

The second route (the Popperian route) through the scientific method clearly recognizes that scientists have preconceived ideas derived from existing theory. Drawing from available theory, a new idea is postulated or put forward – this is a hypothesis. A hypothesis must be stated in a manner that permits falsification. If falsified when tested, the hypothesis obviously requires modification, but if it is not falsified the test may be regarded as "corroborating" the research hypothesis. As will be seen shortly, it is important not to regard the hypothesis as "verified", and use of corroboration stems from conscious rejection of the concept of verification (Haines-Young & Petch 1986, pp. 68–71). In passing, it is worth noting the tremendous importance of the hypothesis itself: it must have the potential to become a law; if it is trivial, it is not even a hypothesis. This means that a hypothesis is likely to emerge after some considerable intellectual effort and is not something that appears very early in research merely to get the ball rolling.

It is, perhaps, fairly easy to see the objections to the Baconian route and, conversely, rather easy to accept the Popperian route. However, the latter route is itself a model fraught with intellectual problems, and is the source of enormous debate among historians and philosophers of science. An exhaustive analysis will not be attempted here; the topic is discussed by

Harvey (1969) and has recently been scrutinized, with particular emphasis on physical geography, in an outstanding fashion by Haines-Young & Petch (1986). Only a brief summary appears here, but every geomorphologist should own a well-read copy of Haines-Young & Petch (1986).

Deduction, to postulate something from theory, is widely preferred to induction in scientific circles. However, the most serious objection to it is profoundly basic; if a scientist operates along the Popperian route he cannot investigate anything that does not already have a theoretical foundation. This means, by implication, that theory itself cannot grow and generation of new knowledge is impossible. A second principal difficulty stems from the widely held misconception of verification; how is it possible to subject anything to all relevant tests so that it may be accepted with absolute certainty? Alternatively, how can anything be immune to falsification under all circumstances? Consequently, even when striving to follow the scientific method precisely, the individual scientist must choose between options. The most widely cited conflict is that between Kuhnian and Popperian perspectives.

Kuhn (e.g., Kuhn 1970) and Popper (e.g., Popper 1972) have both published widely and for most purposes it is probably most practicable for a geomorphologist to turn to a summary; Bird (1975) provided one succinct such effort, while Petch & Young (1978), Young & Petch (1978), and Haines-Young & Petch (1980) have pursued the topic at length. This sequence of papers is surpassed by their recent text (Haines-Young & Petch 1986) in which both this issue and several other important aspects of methodology are reviewed succinctly, but cogently and comprehensively.

The essential contrast is between Popper's emphasis on falsification, leading to the sequence: $P_1 \rightarrow TS \rightarrow EE \rightarrow P_2$, where P_1 = a problem, TS = a tentative solution, EE = error elimination, and P_2 = a second problem derived from TS as limited by EE (Bird 1975, p. 153). This simplified sequence is usually reiterated as in Figure 2.4b. The simplified sequence may be contrasted with a second one based on Kuhn's perspective: $P_1 \rightarrow TS \rightarrow V \rightarrow T$, where P_1 = a problem, TS = a tentative solution, V = a verification procedure that either eliminates, modifies, or accepts TS, and T = a theory, law or paradigm (Bird 1975, p. 154).

Kuhn has presented a description of how the scientific community seems to operate. A governing code (paradigm) is accepted and the community goes about its routine, daily business for lengthy periods. Infrequently, dissatisfaction with the prevailing paradigm and/or the buildup of un-answerable questions reaches a level that leads to convulsion (scientific revolution). At this time the prevailing paradigm is discarded and instantly replaced by one offering more appeal and scope.

Popper offered a more idealized or naive view of science in which scientists daily accept the futility of ever proving anything, but constantly seek to disprove existing theory by subjecting it to tests designed to falsify it.

Popper's heavy reliance on deduction lays him open to the basic question confronting all exclusively deductive approaches: how is new knowledge created? Popper suggested that verification is a futile concept; Kuhn countered that all theory is ultimately falsifiable. Therefore, falsification (like verification) is really a relative, not an absolute, concept.

To the reader without a training in philosophy, the ultimate contrast between Kuhn and Popper seems to be that the former described the way it is and the latter the way it should be. In addition, Kuhn tried to say something about the way science is created, while Popper dismissed this as irrelevant and indicated how science should be processed and presented. Quintessentially, neither falsification nor verification is entirely without uncertainty, but Haines-Young & Petch (1986) present a convincing argument in support of Popper's concept of falsification. Nevertheless, even if this is the way we should operate, it is clearly not the way most researchers generally do, because research journals are quite obviously dominated by papers in which the tenor of presentation is one of displaying proven or verified results.

Some admonitions, caveats, and options

There is no substitute for a well-defined and fully grasped objective. A disciplinary philosophy is not an unnecessary or fanciful accoutrement; rather, it is a guiding light without which all technique is unnecessary and fanciful. On this matter there would appear to be no option if a quality product is to ensue.

Once girded with an objective, selection of a theoretical perspective is the most important methodological step to be taken. Not only should this be done initially, it should be done consciously, for as Chorley (1978, p. 1) put it " . . . the only true prisoners of theory are those who are unaware of it". In fact, not only is theory not a prison, it is the very medium through which human beings address their objectives. An extract from Margenau (1961, pp. 28–9) highlights the issue:

Yet it is evident to every scientist that a bare and isolated fact, an unrelated experience, a single observation, commands no one's sustained attention; it is inherent in the nature of an isolated fact to be *un*satisfying, to clamor for context and fulfillment. But a set of *related* and *suggestive* facts, a significant experience, a set of observations called an *experiment* – these are the building blocks of science and of philosophy of science. And what is it that makes a fact related to other facts, makes a perception suggestive, an experience significant, a set of observations an experiment? Clearly they require a certain background of interpretation to take on meaning, a medium of expectations which they confirm or refute, a texture of theory which they illuminate. A forest of facts unordered by concepts and constructive relations may be cherished for its existential appeal, its vividness, or its nausea; yet it is meaningless and cognitively unavailing unless it be organized by reason. Reflection upon this matter leads to this somewhat paradoxical and startling conclusion: facts are not interesting or important ingredients of science unless they point to relations, unless they suggest ideas combined into what is called theory.

Within geomorphology there are a number of possible objectives, to be discussed shortly and therefore bypassed here. Whichever the geomorphologist selects, he or she must consciously choose an appropriate body of theory, for it is improvement of theory that advances understanding or explanation. Functionally, a geomorphologist is most likely to use the scientific method and, although there are other options, it is to this method that the following remarks are limited.

If the scientific method is to be used, the first requirement is the ability to distinguish scientific from nonscientific. In the critical rationalist, or Popperian, view, the specific test is that of falsifiability. That which is falsifiable is scientific. Once into a scientific mode in the general sense it becomes necessary to know how to operate in more specific terms.

Haines-Young & Petch (1983) suggested that the multiple-working-hypothesis approach, first formalized by Chamberlin (1897) and expanded by Johnson (1933), still represents the best approach. A cursory inspection will not reveal any self-evident compatibility between the multiple-working-hypothesis approach and critical rationalism, but the two are compatible providing that some distinctions are made rather carefully.

Chamberlin (1897) was concerned with narrowmindedness, associated with what he called the "ruling theory" approach to research. He characterized this (Chamberlin 1897, p. 841) as premature explanation becoming tentative theory, then adopted theory, and, finally, ruling theory. While championing the multiple-working-hypothesis approach, he recognized the difficulty of expressing parallel and complex ideas in the written word and mathematically. Such difficulty arises from the fact that both languages are necessarily linear or sequential.

Chamberlin's ideas were consolidated in a detailed re-examination by Johnson (1933). Seven stages of investigation were recognized by Johnson: (1) observation; (2) classification; (3) generalization; (4) invention; (5) verification and elimination; (6) confirmation and revision; (7) interpretation. At all stages Johnson urged the importance of "analysis", which he defined as "the process of separating observations, arguments, and conclusions into their constituent parts, tracing each part back to its source and testing its validity, for the purpose of clarifying and perfecting knowledge" (Johnson 1933, p. 469).

Emphasis on observation, classification, and generalization are clearly inductive and would seem to be at odds with critical rationalism. Similarly, the term "verification" does little, at first glance, to square with the concept of falsification. However, these apparent conflicts between multiple working hypotheses and critical rationalism may be resolved. Stages (1) through (4) are irrelevant to critical rationalism, because Popper argued that the method of hypothesis formulation is irrelevant to the status of the hypothesis as scientific or nonscientific. Therefore, steps (1) through (4) may simply be called good practical advice.

Verification clearly seems to imply the opposite of falsification, yet Johnson (1933, pp. 482–6) emphasized the importance of deduction at this stage of research and also emphasized that the researcher can only conclude that a verified hypothesis is "competent" to explain the situation, and cannot conclude that he or she has found the "true" answer. In this sense the tenor of Johnson's ideas is compatible with critical rationalism and his use of the term "verification" is misleading when compared to its usual use in this context. In nearly a century of thought, geomorphologists seem to have moved along many circuitous paths, split many a hair very nicely, only to remain pretty much where we were!

In essence it seems that rules cannot be legislated for the magic moment of creation or insight. The scientist's mental contortions, gyrations, or occasional elegance are simply too diverse and obscure to be specified. Therefore, it seems practicable to accept that science refers only to the final and formal presentation of research. Conversely, as Chamberlin (G. K. Gilbert before him) and Johnson realized, it is perfectly possible to follow a strategy designed to produce as efficiently as possible the best resource base from which to create science. The multiple-working-hypothesis approach is one good system – being a genius is equally acceptable.

Regardless of the manner in which it is formulated, the hypothesis is clearly the linchpin of science. Implicit in the term "hypothesis" are the notions: deduced from theory; postulation of a substantive question; potential for making a law-like contribution; falsifiable (i.e., measurable). Phrased rather more colloquially, everything depends on the quality of the question. A good question is pithy, it reflects the researcher's grasp of existing theory and where it might lead, and it also poses something manageable and creates the possibility of a precise and insightful answer. In contrast, a poor question lacks one or more of these attributes.

The diversity and sophistication of the contemporary technical arsenal is another two-edged weapon. Unquestionably, it has permitted enormous expansion of scientific horizons (which are conceptually infinite), but in many instances technical sophistication has been used as a substitute for a sound theoretical foundation. Neither the computer nor the scanning electron microscope, nor the canonical analysis nor magnification they produce, is a satisfactory substitute for a sound hypothesis; however, they may, and frequently do, make an outstanding complement.

The purpose of this chapter has been to illustrate the nine-tenths of the scientific iceberg that scientists in Earth or field sciences often prefer to ignore. In reality, theory so permeates science that to ignore it is analogous to ignoring oxygen on the grounds that breathing is a reflexive body process. Unfortunately, science is a man-made artifact and contains no reflexive processes that function untended.

Two complementary issues must now be addressed more fully. First, it is necessary to examine the possible objectives open to geomorphologists or to answer the question: "How many geomorphologies are there?" Second, thinking is not only channeled by grandiose theoretical notions, but also by the most basic component of daily thought and interaction – words. The technical vocabulary, or jargon, of a discipline is a vital shorthand for intradisciplinary communication. Consequently, its use and misuse merit sharp scrutiny.

3 Geomorphology – definitions and approaches

What is geomorphology?

If we are to answer the question "What is geomorphology?", it is necessary to establish both the objectives of geomorphologists and the domain (subject matter) of the discipline. The word "geomorphology" comes from the Greek stems *geo* (earth), *morphos* (shape), and *logos* (reason); hence, the study of the Earth's shape. Defined in this fashion the term, which was first used in English in the late 1880s (see Tinkler (1985, p. 4) for a brief discussion), appears to prescribe both objects and domain. However, even a cursory review of contemporary research geomorphology reveals that it incorporates investigations of only highly selective fragments of material that one would anticipate from etymology. At the largest scales, the Earth's initial evolution falls into the fields of astronomy and geophysics, while its overall shape is the domain of geodesy.

A second set of spatial and temporal scales may be envisaged at a continental level, where continents are viewed as basic spatial components and timespans are sufficiently lengthy to identify changes in continental configuration and/or location. Most people recognize this as the realm of plate tectonics, a powerful theoretical framework (paradigm for some!) for studies of the Earth's crustal behavior. In general, geomorphologists contribute little to plate-tectonic research and it is accepted as the domain of geophysicists and geologists.

Somewhere, admittedly poorly defined, at the level of portions of continents (i.e., regions) and periods shorter than the entire geologic record, the first traces of recent and contemporary geomorphology emerge. Geomorphic research undertaken in the recent past at this regional scale is exemplified by physiography (e.g., Fenneman 1931, 1938), regional geomorphology (e.g., Thornbury 1965), and denudation chronology (e.g., Johnson 1931, Wooldridge & Linton 1955). It is most strongly represented in current geomorphology by continental European climatic geomorphology (e.g., Tricart & Cailleux 1972), but also exists in a limited fashion among English-speaking geomorphologists as mega-geomorphology (Gardner & Scoging 1983). At much smaller scales, the heart of contemporary geomorphology in the English-speaking world emerges in various guises, commonly expressed as process, quantitative, theoretical, geotechnical or applied geomorphology.

Geomorphologists not only usually restrict themselves in terms of scale alone, but also do so with respect to subject matter. At any moment in time, configuration of any point or area on the Earth's surface is the net result of endogenetic (internal) and exogenetic (external) forces. However, although geomorphologists occasionally make contributions to endogenetic research (incorporating tectonic, structural, and volcanic processes), they are overwhelmingly consumers of such information.

Exogenetic processes, embracing sub-aerial and sub-aqueous weathering, transport, and deposition, are the broad domain within which the majority of geomorphologists operate. In reality, even this sphere is not fully occupied or wholly unchallenged. Sub-aqueous research in fluvial and lacustrine contexts is largely geomorphological, but submarine research is poorly developed and probably dominated by those most closely identifying themselves with oceanography. Fluvial and coastal deposition are clearly part of geomorphology, but geomorphologists investigating them quickly find themselves amidst sedimentologists and stratigraphers; and, increasingly, Quaternarists.

Given the particular needs of this text it is necessary to struggle a little harder to define the issues more clearly, because without well-defined objectives there would appear to be no identifiable discipline. Therefore, the objectives of geomorphology must be pursued more closely. Accepting that such objectives are likely to have evolved and that geomorphologists are probably divided on their objectives, a brief historical review of the main groups within the discipline is appropriate.

Some historical objectives in geomorphology

A comprehensive review of the development of geomorphology, up to and including the life of William Morris Davis, is to be found in Chorley *et al.* (1964, 1973). While this is fascinating reading, these two large volumes represent a sizable undertaking. Tinkler (1985) provided a much briefer review that treats the distant past less thoroughly, but reaches nearer to the present day than do Chorley *et al.* While Tinkler's presentation has some limitations, as would any individual's brief review of a large topic, it does provide an informative summary of the history of geomorphology that may be readily absorbed. A third source, the briefest of the three, is provided by Chorley's (1963, 1978) papers.

Chorley (1978, p. 3) pointed out that for several hundred years geomorphology was viewed as an inherent expression of God's will. Such a philosophy was lost only in the early 19th century when the work of Hutton and Playfair (and later that of Powell and others) was sufficiently widely accepted to provide an alternative foundation for the discipline; at this time early teleological philosophy was replaced by an immanent one. While it is usual to view Hutton, Playfair, and Powell as those first bringing a truly "scientific" approach to geomorphology, Chorley's choice of immanent philosophy to describe their perspective is an attempt to convey something extra about the character of their views. "Immanent" may be regarded as a synonym for "intrinsic" or "inherent". Therefore, these early workers are characterized by their ideas of inbuilt balance or interdependence between natural phenomena, e.g., the inherent balance between erosion and deposition, or the intrinsic relationship between bedrock type and topography. As Chorley pointed out, the great success of these views derived from the large scales used by these early researchers; as detailed studies appeared, so did discrepancies and, finally, dissatisfaction.

Another objective that has been important in geomorphology and would seem to be an obvious derivative of strong ties with geology is reconstruction of regional landscape evolution. This is well-exemplified by denudation chronology, i.e, reconstruction of the landscape in a sequence of stages, each with a specified age (e.g., in Britain, Wooldridge & Linton's (1955) evolution of the Weald; in the U.S.A., Johnson's (1931) version of Appalachian development). The origins of denudation chronology pre-date Davis's geographical cycle, but the latter was so compatible with the former that after its introduction it soon became the theoretical underpinning for denudation chronology and the two became inextricably entwined. Ultimately, this approach was undermined by geologic time whose span was too great for available dating techniques and, therefore, reduced denudation chronology to a series of fanciful re-creations devoid of the necessary factual evidence to convince "scientific" geomorphologists.

The final pre-World War II philosophical grouping recognized by Chorley is a functional one, stemming from the surge in information derived from European colonization, as well as from internal exploration within the U.S.S.R. and the U.S.A. This trend is most readily apparent in climatic and morphological geomorphology, both of which are still recognizable entities within contemporary geomorphology. So at this point it is practicable to view contemporary, or post-World War II, foci in geomorphology.

Contemporary geomorphology

It is not possible to categorize the activities of each and every geomorphologist, but it is possible to identify some rather obvious and large groups that, in sum, embrace the majority of geomorphologists. What follows is a series of thumbnail sketches of principal groups.

Morphology

Implicit to an examination of morphology or shape is the notion that categorization (classification or taxonomy) will provide insight into evolution (see Haines-Young & Petch (1986, pp. 158–63) for some interesting comments on this). Much modern morphology stems from the work of R. E. Horton, an American hydrologist; although it is A. N. Strahler who is really responsible for developing the potential created by Horton's pioneer work.

Form was examined in its most general two-dimensional (slope profile) and three-dimensional (drainage basin) manifestations in Strahler's (1950, 1952b) papers, respectively. Some other classic or benchmark papers quickly followed; perhaps none was more influential than Schumm's (1956) paper on badland evolution. Schumm's paper helped to establish in the minds of geomorphologists everywhere the notion that badlands might serve as the critical missing link between the laboratory and the real world. A notable exploitation of this idea was Parker's (1977) work on large, artificial network experiments; however, several papers in Bryan & Yair (1982) cast doubt on just how valid this view of badlands may be.

Chorley (1978, p. 8) labeled the approach of Strahler and his coworkers "functional", pointing to dependence upon statistical analysis. It is also worth noting the importance of scale-free (ratio) assumptions that played a large role in making extrapolation seem feasible. Such approaches to geomorphology are obviously directed to generalization of evolutionary principles, but provide no insight into the timespans involved in the real world.

Morphology not only lends itself to analysis of field-generated data, but is an attractive approach for those interested in manipulating data derived from topographic maps, aerial photographs, and more recently from digital tapes. Given the so-called "quantitative revolution" in geography, it is not surprising to find that many geographically trained geomorphologists indulged in highly statistical morphological investigations; Chorley (1972) provided a representative selection of the directions in which morphological studies diverged, while Evans (1972) is a good summary of the basic concepts involved. In a more recent review, Gardiner & Park (1978) once more emphasized divergence in morphological studies as researchers pursued ever more sophisticated techniques.

It is appropriate to remember that remote sensing has emerged as a potentially profound contributor to this sphere of geomorphology (e.g., Verstappen 1977). Regular, repetitive coverage by satellite would appear to offer an enormous step forward if morphological change could be detected. Progress in this area appears to be hampered by data management (simply sorting and manipulating a veritable torrent of data) and sensor resolution (an academic, rather than a military, problem!). At present, commonly available sensor resolution does not match the scale of most process research.

Climatic geomorphology

Climatic geomorphology rests on the idea that there are recognizable suites of landforms that may not only be identified, but clearly established, as products of particular climatic regimes. While this has not been a particularly popular approach in the English-speaking world, its importance in French and German geomorphology is profound (e.g., Tricart & Cailleux 1972, Büdel 1982). Implicit is the linkage: climate controls process(es) and process(es) control form; therefore, form is a product of climate.

Most original climatic geomorphology has been published in French and German, an exception being Peltier's (1950) German-inspired paper. A reasonably good feel for the tenor of such work may be derived from various translations (e.g., Derbyshire 1973, Tricart & Cailleux 1972, Büdel 1982) as well as from reviews such as that of Stoddart (1969). Another volume edited by Derbyshire (1976) is far from the mainstream of climatic geomorphology and is appropriately characterized as a collection of essays by process geomorphologists attempting to evaluate climatic geomorphology from a purely process perspective.

Chorley (1978, p. 6) placed climatic geomorphology into a taxonomically based group, and certainly climatic geomorphology is infused with morphological studies and very short on process research. Chorley also went on to note the theoretical shallowness of climatic geomorphology and the chameleon-like qualities it has exhibited as other philosophies have waxed and waned.

Process geomorphology

R. E. Horton, whose work did so much to stimulate early morphological studies, was also an important source of inspiration for early process geomorphologists. Process geomorphology was characterized by Chorley (1978, p. 7) as being overwhelmingly mesoscale, and functional in philosophy. Transfer of water and surficial debris serve as surrogate (substitute) measures of processes themselves; hence such terminology as overland flow, sediment discharge, and rotational slump.

Analytically, process geomorphology has always been dominated by statistical procedures. Theoretically, process rates and their spatial and temporal variability have been central. Critical conceptual underpinnings include magnitude and frequency and the ergodic hypothesis. In the field, rapidly changing environments such as beaches, badlands, and some periglacial regimes have been particularly popular subject areas. At present, process geomorphology seems to be most severely restricted by problems emanating from scale-linkage issues.

Geotechnical science

A more recent offshoot from process geomorphology is geotechnical science, or phrased more colloquially "geomorphology from an engineering perspective". This approach has large theoretical infusions from civil and soils engineering, as well as from what is termed "materials science". A quotation from a recent text by Williams (1982, p. vi) serves to convey the geotechnical perspective:

> The explanation of what we observe in the natural environment immediately around us, as well as globally, lies in the behaviour of the individual gaseous, liquid or solid components of the earth's surface. Axiomatic though this may seem, most books in physical geography, and other fields pertaining to the environment, concentrate instead upon the composite characteristics of localities or regions. Yet landscapes, terrain or climate represent the interaction of many different phenomena and materials, and this is also true of quite small features – a small area of ground surface or a particular hillslope, for example. This book is concerned with understanding the natural environment through a detailed examination of what may be called, by analogy with "microclimate", the "microenvironment". This involves consideration of specific material properties and processes.

The theoretical underpinning here is fairly obvious; geomorphology may be appropriately viewed as the chemical and physical behavior of surficial Earth materials. While such a perspective provides a sound starting point, successful linkage from molecular to regional scales is going to be a formidable task. Nevertheless, there are some extremely provocative texts that suffice to highlight what can be achieved, among the most informative being Yatsu (1966), Scheidegger (1970), Johnson (1970), Carson (1971), Whalley (1976), and Williams (1982). For those without a good science background, Davidson's (1978) summary of basic science will serve as excellent preparatory reading for any of the preceding texts.

The theoretical underpinning of geotechnical science may be very strong in site-specific studies. However, when attempts are made to generalize data that have probably been generated in laboratory analyses, scale linkage quickly becomes a major dificulty (e.g., Chorley 1978, p. 10). At present, it is too early to tell what this realist philosophical approach will contribute to geomorphology, but its growth is certainly associated with that of applied, as well as process, geomorphology.

Applied geomorphology

Applied geomorphology overlaps with process geomorphology and geo-technical science, as well as emerging in portions of Earth science, environmental science, hazard perception, and soil conservation – to name only a few of the more obvious areas. Stemming from a conventionalist philosophy (Chorley 1978, p. 10), applied geomorphology may be seen as being driven by the widespread urge to make all disciplines relevant. Without making too much of the issue of relevance, it is easy to see that research monies and associated prestige are most readily available to those whose work is of immediate use to society, regardless of academic discipline. However, it is important to appreciate that there is nothing inherently good or bad about applied geomorphology. There has been a good deal of rather poor regionalization (classification) undertaken in the name of applied geo-morphology, but this may be contrasted with some truly excellent geo-morphic research pursued in the same vein. Research conducted by Australia's Commonwealth Scientific and Industrial Research Organization (C.S.I.R.O.) is a prime example of the latter (e.g., Christian & Stewart 1953).

Applied geomorphology has led those geomorphologists pursuing it into close working relationships with engineers. In and of itself this is not a daunting problem: engineers have many techniques that are highly appropriate to site-specific studies, while geomorphologists may make significant contributions to projects embracing large spatial scales. The primary difficulty with such collaborative work is that it is almost exclusively devoted to human timescales. Thus, the primary intellectual issue appears to be that geomorphologists are being drawn away from timescales at which landscape evolution occurs, that is, they are ceasing to function as geomorphologists. Again this may be summarized as the scale-linkage problem in a slightly different form.

Geomorphology or geomorphologies?

To answer the question posed above in the singular, it must be possible to identify a single (set of) objective(s) and preferably, but not necessarily, a single body of theory. It is also necessary to establish that there are objective(s) and/or a domain unique to the discipline. If a single (set of) objective(s) can be identified, but diverse bodies of theory are used, it is necessary to recognize the theoretical diversity and evaluate whether it is wholly divergent or may be considered potentially convergent.

Traditionally, the core of geomorphology has been mesoscale. Spatially this has meant regional studies, something larger than individual slopes, but smaller than complete continents. Temporally it has embraced periods much longer than human lifetimes, but very much shorter than the entire geologic record. The domain of geomorphology has always been focused upon landscape change, but nearly always with explanation of past evolution, rather than prediction of future change.

Identification of a core or focus is important, but there have always been significant peripheral research interests that have served to enrich the mainstream. One example is the contribution made by Walter Penck; his own interests were centered on endogenetic processes, but his use of exogenetic landscape forms to pursue his goal resulted in important contributions to geomorphology. Examination of relationships between endogenetic processes and landscape form has remained an important secondary interest of many geomorphologists; Morisawa & Hack (1984) reflects the recent resurgence of interest in endogenetic processes and landscape form. Research into basic geomorphic processes also has a long history (see Tinkler (1985) for some brief illustrations, or Chorley et al. (1964) for a comprehensive treatment) and many contemporary U.S. geomorphologists have drawn their inspiration directly from the work of G. K. Gilbert, thereby placing much greater emphasis on the contemporary balance between form and process and reducing emphasis on the past.

In fact, dissatisfaction with Davis's geographical cycle (a regional approach) and seeds sown by Gilbert's largely process-oriented approach played a major role in the growth of modern process geomorphology. However, during its growth, process geomorphology has exhibited a major shift in emphasis with researchers tending to work their way down the spatial scale (looking at ever smaller issues) rather than retaining their interest in mesoscale issues. It appears that such a shift has produced a firmer theoretical basis for the research undertaken (because physics and chemistry have become the foundation), but the original objective of explaining landscape change has either been consciously abandoned or inadvertently misplaced.

At present, it appears that geomorphology is splitting or in metamorphosis. Dissatisfaction with Davis's geographical cycle stimulated process geomorphology, which was originally taken up in the belief that it would provide more satisfactory answers to existing problems, i.e., mesoscale issues. However, as it has developed, process geomorphology has actually ended up addressing a different set of problems, i.e., microscale issues. While some process geomorphology may still be related to regional-scale research, the bulk of it is not. Furthermore, process geomorphology has spawned geotechnical research, which generally functions at an even more fundamental level.

Meanwhile, there has been little or no theoretical growth in regional or mesoscale geomorphology. Climatic geomorphology appears very superficial and poorly founded to those with any process training. L. C. King in South Africa has tried to create an alternative theoretical framework to

Davis, but, while his work has some serious flaws, its merits have received much too little attention. Only a very limited number of researchers have attempted to wed process and regional geomorphology. Twidale (e.g., Twidale 1976, 1983), working in Australia, is one person who has pursued this combination by matching traditional fieldwork with laboratory analyses, while also consciously considering relevant theoretical issues. Kirkby (e.g., Kirkby 1971, 1986) has steadfastly pursued a more radical approach by building a mathematical model that started at the scale of the individual hillslope and is now emerging at a larger scale.

Another fundamental shift associated with development of microscale process and geotechnical research is conscious concern with prediction, rather than with traditional "retrodiction". The primary difficulty with this, at present, is that the inputs are of a kind that permit modeling of sediment transfer and similar issues, but are not suitable for prediction of long-term landscape evolution.

Many national schools of geomorphology in the non-English-speaking world appear to have retained greater homogeneity than their English-speaking counterparts. Be this as it may, contemporary English-speaking geomorphology shows considerable diversity, which may be interpreted either as metamorphosis in progress (i.e., the diversity represents a mixture of the old and the new with the former dying and the latter growing) or as fundamental divergence (i.e., the old goals and discipline remain vigorous while new (sub)disciplines emerge and pursue their own aims). This presents certain obvious difficulties in reviewing relevant theory.

One viewpoint is to declare the old geomorphology, or one of the new geomorphologies, valid and everything else invalid. This opens the way to examination of a restricted, but highly focused, body of theory. An alternative approach is to accept some discrepancies and to examine a more broadly based body of theory. The latter approach is adopted here. It is done for two reasons: (1) this text is addressed to students, many of whom may not be aware of many of the relevant (let alone tangential) theoretical issues; (2) the divergence in geomorphology that has been suggested here has been happenstance rather than conscious. In fact, the second point may be amplified to highlight the fact that it is the widespread neglect of theory in geomorphology that is fueling divergence. One of the primary purposes at hand is to suggest that greater familiarity with, and attention to, theory would create a stronger and more unified discipline.

So with some foci highlighted, some weaknesses exposed, and some strengths sketched, it is almost time to face the objective of this text – namely a comprehensive, but not necessarily exhaustive, survey of theory underpinning contemporary geomorphology. The final preparatory step is at once trivial and profound, an appraisal of the vocabulary used in geomorphology.

4 The importance of terminology

Introduction

In the Baconian view of science, facts exist independently of science itself. Consequently, when they are gathered and ordered to produce science, they do not themselves influence the scientific process. Traditional views of the relationship between words and thinking were very similar to this. Words were considered to exist independently of thought itself, and when organized to express thought they themselves were not considered to influence the thinking process. Relationships between thoughts and words are studied in several different contexts, two obvious ones being philosophy of science and linguistics. In the latter field, the traditional word–idea relationship was challenged by Sapir (1921) and even more strongly and successfully by his student and associate, Whorf (1939, 1940a, b, 1941).

The Sapir–Whorf view of language may be called one of "linguistic relativity" where "the structure of a human being's language influences the manner in which he understands reality and behaves with respect to it" (Carroll 1966, p. 23). The thrust of this argument is that the content and structure of a person's vocabulary play an important (but not an exclusive) role in the way the person thinks. As Carroll (1966, p. 26) noted, this argument is founded on the concept that the content of thought actually influences the process of thought.

Words have meaning and significance as the result of convention; a group of people agree that a particular word will have a specific meaning. This is quite easy to understand when limited to naming objects or classes of objects. For example "apple" is not likely to be a word that generates much debate, nor does it provide much problem when translated into French (*pomme*) or German (*Apfel*). However, it is very obvious that an enormous number of words have very much more complex meanings than this simple example. If a word has a meaning it may be viewed as a symbol (i.e., it stands for or represents something). As a conventional symbol, a word must have a definition and any definition always leads back to theory.

The strength of natural (everyday) language is its flexibility. It is something that has grown and changed, and continues to grow and change, to meet the needs of the times. One of its greatest strengths is that there are

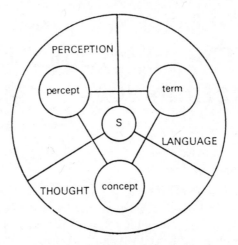

Figure 4.1 Relationships between percepts, concepts, and terms (after Caws 1965). (From Harvey (1969).)

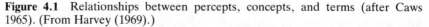

many ways of saying the same thing. Given such strengths, it seems very likely that definition, which requires great precision in the scientific world, is likely to be a very difficult issue. This is partially revealed in dictionaries, which do not actually define words, but provide a synonym or set of synonyms that permit the reader to use the word appropriately; furthermore, they confound definition by illustrating that many words have multiple meanings. Finally, many dictionaries reveal to the reader how the meaning of a word has changed over time. In fact, this web of words bedded in culture and history is so complex that mathematicians and logicians have essentially declared it unmanageable and resorted to a variety of artificial languages. However, those specifically interested in language itself have continued to struggle with the meaning of words.

In trying to systematize how words are handled, it is common to identify three parts. For example, Caws (1965, pp. 31–6) used: sign (in this context, a word); object (the thing for which the sign stands); the interpreter (a person). In Ogden & Richards (1949), the three parts are symbol (word), referent ("designata", or what is referred to), and reference (thought, approximately the same as the interpreter). The general pattern then is that people sense things called "perceptions", think things called "concepts", and speak or write things called "words". The overriding problem is that there is not a one-to-one transfer between perceptions, concepts, and words. Harvey (1969, p. 19) provided a schematic diagram of Caws's view of these difficulties (Fig. 4.1). A simple illustration of the problem is that there are both perceptions and concepts for which there are no words (terms).

The real crux of the Sapir–Whorf argument may now be visualized as a feedback relationship. If there is not perfect correspondence between the three sets in Figure 4.1, how will thinking be constrained by lack of appropriate words? How will people communicate thoughts for which there are no words? Whorf made much of his case by comparing the Hopi and English languages. The immediate need here is to recast the argument in geomorphic terms.

Terminology in geomorphology

The terminology of a discipline is in large part a shorthand, as well as a language. As with any language, a group of people have agreed, by convention, to accept a set of terms; in this sense scientific terminology faces problems that occur in all languages. However, scientific terminology is a particularly intense yet stark language type: intense because many abstract and complex ideas are involved; stark because great precision in definition is needed. Thus, under ideal circumstances, the terminology or jargon of a discipline permits clear communication devoid of laborious descriptions by virtue of appropriately used terms. The negative side of scientific terminology is again no different in principle from any other language, merely a heightened version of it. Possible flaws in the system are that: (1) a term may not, in fact, be well-defined; (2) a term may be misused by the speaker; (3) a term may be misunderstood by the listener; (4) a term may have more than one definition; (5) a term may have evolved.

In actuality, there is yet another possible source of difficulty – the word may be meaningless to the listener. If words are defined with respect to a body of theory, it may be that a particular term describes a nonexistent concept if the listener does not subscribe to the same theoretical perspective as the speaker. In this sense each theoretical view of geomorphology has created its own language. Accepting the Sapir-Whorf notion that vocabulary influences thought, it is relatively easy to see how adherents of various schools of geomorphology channelize or reinforce their views by virtue of their vocabulary.

However, inasmuch as terminology is imbedded in theory, it is possible to see that geomorphic terminology has suffered the same benign neglect that has plagued theory itself. One first step in attempting to rectify the situation is to consider what is needed for geomorphologists to communicate with each other on a regular basis.

Description and interpretation

Prior to any other activity a researcher must declare his theoretical position because it constrains all his scientific activities. Within this constraint the most common activities are description and interpretation. Assuming that two researchers are functioning within the same theoretical framework, it is much more likely that they will agree on the description of something than they will on the interpretation of it. This is not surprising because description, given theoretical unanimity, is a relatively simple task. Therefore, it is desirable to present observations, descriptions, and measurements in a fashion entirely devoid of interpretation. In fact, this is a very difficult task because the technical vocabulary in geomorphology is woefully weak in purely descriptive terms. Furthermore, the pressure on research journals is now so great that editors are very reluctant to allocate space to untreated data presentation.

Ideally, interpretation will commence with presentation of a hypothesis stemming from appropriate theoretical considerations. It is this theoretical structure against which the descriptive material will be evaluated. Again this step is often mishandled because researchers fail to articulate their theoretical perspective adequately, but it is a step that is also poorly served by terminology. This is because the geomorphic technical vocabulary is rich in terms that combine both genetic (causal) and descriptive elements. When description and genesis of a landform are both embraced in a single term, it becomes impossible to discuss either component separately without using a set of qualifying statements – it is precisely these lengthy descriptors and qualifiers that terminology is designed to avoid. A few examples will demonstrate the nature of the problem.

Terminological problems

Morphogenetic terms

A morphogenetic term not only describes the way a landform looks, but also purports to explain how it was formed. A classic example of such a term is "peneplain"; this is not only a surface of very gentle relief but quite specifically a surface that emerges during the penultimate stage of Davis's geographical cycle (Ch. 10). In this particular instance the situation is worsened by the possibility of spelling the word "peneplain" or "peneplane" and thereby making a basic change in definition (Boyer 1979). A synonym for the purely descriptive part of peneplain is erosion surface (although one had better be prepared to show that such a surface is not structurally controlled!). A portion of the genetic connotation may be conveyed by the adjectives sub-aerial, fluvial, marine, glacial, or even polygenetic. However, there is no synonym for the genetic component of peneplain that implies its sequential position in a Davisian cycle.

Another widely used term is "till", an unsorted deposit of glacial origin. Under many circumstances this morphogenetic term may be used without much difficulty. Yet there are a multitude of ancient deposits and reworked coastal deposits that have been labeled till in either a foolhardy or heroic fashion. Again it is possible to separate the constituent elements by a careful choice of words. A diamicton (Flint *et al.* 1960, Washburn *et al.* 1963) is a poorly sorted deposit, regardless of origin. If an origin can be successfully determined, the diamicton may be described as being of glacial, fluvial, colluvial, or some other origin (see, e.g., Madole 1982).

Inadequately defined terms

There are a number of entrenched terms that, upon close inspection, are inadequately defined. Nivation (Matthes 1900, Thorn 1988) is one example; it is a collective noun that embraces an uncertain number of individual weathering and transport processes, plus a specific notion of intensification. Thus, the basic concept is that weathering and transport are accelerated or intensified beneath and/or around a seasonal snow patch. A checklist (Table 4.1) of the supposed constituent elements of nivation reveals just how uncertain the term really is. The real problem is that the term embraces too many separate ideas, and this in itself makes definition difficult. However, further difficulties arise because knowledge has expanded considerably since Matthes introduced the term. Thus, while nivation is a widely used term in periglacial geomorphology, it really does not have a single universal definition, and its main strength today is that its imprecise definition permits it to mean many different things to different people.

The issue of nivation may be taken a step further by considering nivation hollows, supposedly depressions excavated by nivation. These are most frequently "established" by observation of contemporary snow patches occupying hollows. As snow will accumulate in any leeward depression, such observations hardly define hollow origin. In fact, this is a nice illustration of the limitation of correlation; a correlation itself does nothing to distinguish between cause and effect (result).

Another periglacial term of similar vintage to nivation is solifluction. Andersson's (1906, p. 95) original definition reads: "the slow flowing from higher to lower grounds of masses of waste saturated with water". Andersson's work was undertaken in a periglacial climate, and accordingly some subsequent researchers have argued that saturation in periglacial regimes is usually the byproduct of a frozen subsurface and as such this is implicit to the definition. Other workers have argued that there are other ways to generate saturation (e.g., extremely heavy rainfall) and, as Andersson did not specify a frozen subsurface solifluction may occur in any environment and whenever saturation leads to the prescribed type of movement, regardless of how saturation is created. There have been several attempts to supplant the term "solifluction" (e.g., Dylik 1967). In addition, there have been attempts to refine the definition and concept. Washburn (1980) has been a vigorous proponent of the term "gelifluction" – this is a subset concept, gelifluction is quite specifically solifluction that occurs above a

Table 4.1 Comparison of the attributes of the widely used term "nivation" according to various principal authors.* Note the diversity, conflicts, omissions, and absence of quantitative verification. (From Thorn (1976, p. 1171).)

Topic	Matthes (1900)	Lewis (1936, 1939)	Boch (1946, 1948)	Lyubimov (1967)
Data base				
Extensive observations†	yes (Wyoming)	yes (Iceland)	yes (Ural Mountains)	no (Literature)
Intensive observations†	none	minimal	minimal	none
Quantified contrast‡	none	none	none	none
Snow-patch classification	no	yes	yes	no
Snow-patch profile	text (generalized)	diagram (generalized)	diagram (generalized)	diagram (generalized)
Specifics				
Freeze–thaw cycles, intensified	yes (snow margin)	yes (throughout)	yes (snow margin)	yes (headwall base)
Snow-patch core	protective	intensified activity	protective	protective
Bedrock disruption	not mentioned	intensified	intensified	intensified
Sediment transport	intensified	intensified	intensified	intensified
Principal transport process	not specified	solifluction	solifluction	solifluction
Snowpack	static	static	partially mobile	static
Absence of vegetation	brief comment	comment	comment	extended comment
Chemical activity	not mentioned	not mentioned	intensified	intensified (in karst areas)

  * All authors listed defined nivation as intensification of known processes by a snow patch.
  † Extensive observations involved reconnaissance study of numerous snow patches. Intensive observations are detailed, continuous observations and measurements of one or more snow patches.
  ‡ Quantified contrast requires that a process be measured within the area of snow-patch influence and contrasted directly with the same process elsewhere.

frozen subsurface. Washburn has not been totally successful in establishing his suggested terminology, indeed, gelifluction has even been used with other connotations (Embleton & King 1975, p. 97). So, for some, gelifluction remains an uncertain and obscure term; for others, a synonym for solifluction; and for a few, a specific subset concept.

  Like nivation, solifluction is widely used to describe the genesis of forms, in this instance lobes and terraces. Meticulous field measurements by Washburn (summarized in Washburn (1980)) have shown that frost creep is an important mechanism in the development of some "solifluction" features and may even dominate in some instances. It is not known just how widespread this is, because very few people have even considered the issue. Meanwhile, it is encouraging to see that some researchers appreciate the problem and have carefully separated form and process terminology; for example, Benedict's (1970) use of the form terms turf-banked and stone-banked lobe/terrace to describe features that experience both solifluction and frost creep.

Evolving terms

From a strictly logical point of view it makes sense to create a new term every time a new concept is developed or an old one refined (and thereby redefined). Pragmatism argues against this trend being carried to its logical extreme because of the enormous burden it places on the individual to know an ever-increasing number of terms. As a result of these opposing needs, researchers are very frequently faced with a term whose meaning has evolved through time, or a term whose meaning differs from individual to individual so that its use must always be prefaced by explanation of the manner in which it is being used. The latter situation obviously runs counter to the very purpose of terminology. Morphogenetic terms are particularly prone to being resistant in the literature, while continuously being redefined. This is inevitable because they embrace so much that some portion of their meaning is nearly always being re-evaluated.

As an example we may consider the term "periglacial", which was created by Łoziński (1909) without a formalized definition. Initially, it referred to areas marginal to Pleistocene ice sheets, the climates presumed to have been extant in such localities, and also a very few processes. Like a craton, the term has grown by marginal accretion, new processes and new environments having been added to the periglacial world with little or no formality. Too entrenched to discard, too fuzzy to pin down, it is now one of many terms that embraces so much it tells very little. After all, the dry valleys of Antarctica, the Front Range of Colorado, Point Barrow, Alaska, and the crest of the Cairngorms, Scotland, are all periglacial!

Another interesting illustration of an evolving term and surrounding confusion is Teichert's (1958, p. 2718) scholarly discussion of facies.

Stratigraphic facies was recognized, defined, and named by Gressly in 1838. Before the end of the nineteenth century the concept facies became firmly established, through the works of Mojsisovics, Renevier, and Walther, as referring to the sum of lithologic and paleontologic characteristics of a sedimentary rock from which its origin and the environment of its formation may be inferred. Similar or identical rock types have isopic facies, different rock types heteropic facies. Facies changes must be studied in horizontal as well as vertical direction, with the aim of reconstructing changes of environment in space and time. Genetically interconnected isochronous facies form facies tracts; genetically interconnected facies tracts form facies families. Heteropic facies in vertical succession form facies sequences.

When used to designate major stratigraphic sequences occurring in certain geographic, oceanographic, or inferred tectonic environments, the term facies loses its descriptive objectivity. Non-stratigraphic uses of facies are discussed. Recent American facies terminology is discussed in terms of earlier established European terminology. Facies and biofacies as used in ecology differ importantly from stratigraphic facies and biofacies. Rock characteristics now referred to as "biofacies" in stratigraphy should be called paleontologic facies.

The essence of Teichert's findings was that a restricted descriptive term had been expanded to embrace conceptual content to which it was never intended to be relevant, and beyond that, it had evolved from being a rock description (i.e., something a rock has) to being the name of a rock itself. In short, description and interpretation had been unthinkingly, or at least uncritically, fused, robbing the term of precision in both spheres.

Systematization

A plea for systematization may well seem naive. Such a task would be Herculean, but it is not without precedent – at least on a national scale. Faced with a similar tangle of terminology, the Soil Conservation Service of the U.S. Department of Agriculture (1975) has tried to wipe the slate clean and promote a new terminology called "soil taxonomy". This manual is really devoted to a separation of description and interpretation, plus improvement in the precision of description. Unfortunately, this particular attempt is cursed with a truly cacophonous vocabulary, but the underlying concept is sound. In a similar fashion, stratigraphers try to standardize their terminology, using a committee-and-handbook approach.

Certainly in geomorphology it seems essential at least to take the step of separating descriptive and interpretive terms. While it is desirable to have a new term for every concept, it is impracticable. Therefore, the most practicable step is to develop a more flexible terminology; in turn this means each term will encompass less – but with greater precision. The present overabundance of morphogenetic terms is exactly what is not needed. If the Sapir–Whorf view of the interaction between words and thoughts is valid, and there appears to be no reason to doubt it, thinking is actually constrained by our imprecise terminology. Consequently, the plea for a more rigorously defined terminology is merely one facet of the plea for a stronger theoretical foundation for geomorphology(ies).

5 Uniformitarianism and ergodicity

Introduction

Time and space, plus their interaction, form a nexus or web that permeates geomorphology. While separate treatments of time and space minimize the important interaction component, they are necessary to obtain clarity. Accordingly, time (Ch. 6) and space (Ch. 7) will be treated separately. However, time is so pervasive in geomorphology that it creates the biggest challenges in the discipline, and the struggle to deal with it has resulted in the use of two fundamental concepts, uniformitarianism and ergodicity, which are the themes of this chapter.

Ideally scientists in any discipline would prefer to observe the objects and domain of their discipline in their entirety. Nevertheless, researchers in all disciplines are faced with the problems of being unable to measure some things directly, because they are either too small, too remote, happened too long ago, or are yet to happen. Problems associated with time loom particularly large in geomorphology because of the enormous timespans involved in landscape development and also because of the traditional preoccupation with retrodiction, rather than prediction. Emphasis on retrodiction is undoubtedly inherited from geology, and geomorphology as history is a philosophical thesis that has been developed by Reynaud (1971).

The timespans involved in geomorphology (and geology for that matter) are far greater than can be managed by standard statistical sampling techniques and, while statistics play an important role in geomorphology, their use may be distinguished from the concepts of uniformitarianism and ergodicity. Uniformitarianism was inherited directly from geology, and is much heralded in Earth sciences; ergodicity came from physics (specifically statistical mechanics), and is not only unheralded within geomorphology, but is not even in the vocabulary of many geomorphologists. However, together these two concepts underpin what geomorphologists think they know about the past, and, as such, merit careful scrutiny.

Uniformitarianism

A traditional explanation and definition of uniformitarianism credits Hutton (1788) with introducing the concept, Lyell (1830) with coining the term, and geology with making a profound contribution to science in general (e.g., Longwell & Flint 1962, p. 4).

Definition

There is no single, universally accepted, definition of uniformitarianism; indeed, there is no single, universally accepted, content to the term. The more common versions are (Hubbert 1967, p. 4):

(1) The present is the key to the past.
(2) Former changes of the earth's surface may be explained by reference to causes now in operation.
(3) The history of the earth may be deciphered in terms of present observations, on the assumption that physical and chemical laws are invariant with time.
(4) Not only are physical laws uniform, that is invariant with time, but the events of the geologic past have proceeded at an approximately uniform rate, and have involved the same processes as those which occur at present.

Gould (1967, p. 149) subdivided these issues into two components:

(1) A substantive uniformitarianism, dismissed as untrue, which postulates a uniformity of material conditions or of rates of processes.
(2) A methodological uniformitarianism comprising a set of two procedural assumptions which are basic to historical enquiry in any empirical science. These assumptions are (a) that natural laws are constant in space and time, and (b) that no hypothetical unknown processes be invoked if observed historical results can be explained by presently observable processes.

Clearly, uniformitarianism is either a very fuzzy concept or, indeed, several concepts. However, prior to any attempt to evaluate its contemporary role it is appropriate to know something about its inception and evolution.

Original purpose

Hutton launched his concept into a world steeped in church doctrine where the majority of those interested in geology believed that Earth history encompassed only about 6,000 years. In fact, in one of the most precise pieces of geoscience yet documented, Archbishop James Ussher calculated the time of origin to be 9 a.m., October 26, 4004 B.C. (Chorley *et al.* 1964, p. 13)! The furor and ferment that surrounded establishment of Huttonian–Lyellian uniformity is summarized by Hubbert (1967), and considered at length by Hallam (1983) and Tinkler (1985).

The method forwarded by Hutton was founded on the study of rocks as a source of data (induction!) for a variety of processes including weathering, erosion, transport, deposition, and uplift. He rejected supernatural explanations and sought to establish the enormity of geological time. The fruition of the seed he sowed, namely breaking a claustrophobic and myopic perspective, overshadows completely any present-day complaints. In addition, as with many other sophisticated ideas, much of the original care and many of the original caveats were lost as the concept was proselytized (see Tinkler (1985, p. 82) for an illustration of just how careful Lyell was about stating the uniformitarianism case).

Contemporary uniformitarianism

The determination of some geologists to enshrine uniformitarianism, obvious errors in some versions of it, and the fundamental importance of the issues addressed have combined to maintain an enthusiastic, albeit rather repetitive, debate over the term and its utility. Clearly, any notion that Earth processes have acted with anything approaching constant rates (substantive uniformitarianism) is a misconception, as may be verified by examination of data on topics ranging from plate movement (Press & Siever 1982, p. 446), through uplift and denudation (Schumm 1963b), down to gelifluction (Washburn 1980, p. 204).

In a delightfully erudite and succinct paper, Goodman (1967, pp. 93–4) evaluated uniformitarianism. His comments on substantive uniformitarianism included:

(1)   Past prominence by no means implies present pertinence.
(2)   If the Principle of Uniformity is to be taken seriously, it cannot be identified with any such blatant lie as that Nature remains always the same or moves only with dignity.
(3)   Sometimes we are told that the Principle affirms rather that whatever violent and sweeping changes occur, they are always the result of underlying continuous and gradual processes. But this version can be readily discredited. The picture that physics gives us of the universe is of relative macrocosmic stability overlying violent and discontinuous miscrocosmic activity. The bland and sluggard boulder is a maelstrom of particles dashing madly hither and yon, colliding with each other, flickering into and out of existence.

Nothing favors taking an exactly opposite view in geology. Uniformitarianism of this sort defies the findings and the whole tenor of modern science.

Accordingly, it seems appropriate to view substantive uniformitarianism as a concept that may be put to rest.

Methodological uniformitarianism has been a more resilient concept than its substantive counterpart. It has been attacked in two ways: (1) by suggesting that the natural laws have varied over time (remember a true law is universally valid, i.e., invariant over time and space); (2) by claiming that assumptions embraced by methodological uniformitarianism are so widely used in all sciences that the term is simply superfluous.

Heylmun (1971) made an attack of the first kind, citing a wide range of phenomena that may be expected to vary as the Earth's position in the Galaxy, and that of the Galaxy in the Universe, change. This seems to be an ill-founded commentary as it says nothing about the laws themselves, merely that part of the domain to which they pertain is highly inaccessible and therefore falsification extremely problematic. This would only seem to increase the magnitude of a problem that is already pervasive.

A more telling argument is that presented by Goodman (1967, 96–9), who offered the following:

(1)   Whatever made the world and whatever makes it go, the scientist writes its laws. And whether or not nature behaves according to law depends entirely upon whether we succeed in writing laws that describe its behavior.

(2)   Thus the Principle of Uniformity, construed as denying change or abrupt change in the course of the laws of nature can be rejected as false or futile. The uniformity is not in nature's activities, but in our account of them.

(3)   The scientist demands little of nature but much of himself. Where there is change, he looks for constancy in the rate of change; and failing that, for constancy in the rate of change of the rate of change. And where he finds no constancies, he settles for approximations.

(4)   Efforts to simplify a theory are often thought to be merely for the sake of elegance, but actually simplification is the soul of science. . . . The simplest theory is to be chosen not because it is the most likely to be true, but because it is scientifically the most rewarding among equally likely alternatives. . . . We must not suppose that the principle of simplicity always calls for the choice, among alternative hypotheses, of the one that looks simplest. The injunction is rather to choose the hypothesis that makes for maximum simplicity in the over-all theory – indeed, in the total fabric of science. . . . In conclusion, then, the Principle of Uniform-

ity dissolves into a principle of simplicity that is not peculiar to geology but pervades all science and even daily life.

Consequently, it appears appropriate to conclude, as did Shea (1982) in a comprehensive *post mortem*, that uniformitarianism, as a formally defined concept, lacks any contemporary usefulness. However, it would be short-sighted to ignore a far-reaching informal residue.

Uniformitarianism versus neocatastrophism

Uniformitarianism was originally formulated to overcome both a sup-posedly short geologic record and a widespread willingness to invoke catastrophic, often supernatural, origins for things geological – the biblical flood being the prime example. Lyell was not only a contemporary of Darwin, they were close friends (Tinkler 1985, p. 83). One outcome of the similarity in their scientific ideas was a concept of slow change ("evolution"; see Ch. 10) that became the norm throughout the Earth sciences and natural history. In geomorphology the term "gradualism" is usually used to convey notions stemming from Hutton's uniformitarianism and vigorously pro-moted by Lyell (Garner 1974, pp. 2–3). Gradualism incorporates the ideas that geomorphic change is slow, occurs in small increments, and involves relatively weak forces. As with uniformitarianism itself, it must be remem-bered that gradualism was launched to counter concepts invoking cata-clysmic, and often supernatural, events as the basis for Earth history. One result of gradualism's defeat of catastrophism was that it produced an outlook in which it was very difficult for Earth scientists to accept the true importance of rare, large-magnitude events. This was perhaps an emotional issue as much as a scientific one; anyone who invoked a rare, large-magnitude event as a geomorphic explanation immediately became illogi-cally tainted with the old, supernatural catastrophism. A classic example is the initial response of the scientific community to Bretz's explanation of the Channeled Scablands of eastern Washington State during the 1920s and 1930s (e.g. Bretz, 1923).

Neocatastrophism may be described as a modern reintroduction of the notion that rare, large-magnitude events (even unique events or singulari-ties) appear to have played important roles in geologic and geomorphic history. As a concept it is, of course, devoid of the original constraints of a short geologic record and appeal to supernatural powers. However, it is not always without some notion that past processes are not necessarily fully represented by present processes.

The pressure to re-emphasize catastrophic concepts came initially from paleontologists (Dury 1975, p. 135). Today, paleontologists are inclined to believe that mass extinctions have occurred, and sedimentologists and stratigraphers believe that rocks include a clear record of large, individual events. Several important results of the reappraisal were discussed by Dury (1975, 1980), as well as by Hsü (1986). Geologic mechanisms that favor periods of quiet, interspersed with rare, large events, were enumerated by Parker (1985). In general terms his suggestions amount to no more than standard notions; energy is stored as a reservoir fills, only to be released abruptly as its capacity is exceeded. However, such ideas effectively fill the needs of the situation by explaining timing and magnitude of geological events in a fashion that produces step-function behavior.

There is an important intellectual issue implicit in re-emphasis of the importance of extreme events. If greater significance is now being attached to large events in a series, this represents only adjustment of magnitude and frequency concepts (Ch. 6). However, if the new perspective is one that identifies "unique" events (remember that colloquial U.S. usage of the word "unique" is incorrect – unique means one of a kind, and it cannot be qualified as in fairly unique, quite unique, or very unique; nor does it mean rare) as paramount in geological and geomorphic records, then there can be no sciences of geology or geomorphology, because as Brown (1974, p. 456) noted "there is no science of singularities". In the latter instance, presumably both disciplines are actually a form of history (Reynaud 1971).

It seems that geomorphologists are presently in transition; having escaped uniformitarianism, which served them well but itself became limiting, they have not yet arrived at a new, fully articulated standpoint. The majority of geomorphologists are almost certainly in the position of reappraising magnitude and frequency, and not in that of pursuing singularities. Scientists interested in even longer timescales, and especially those interested in the history of life, face greater problems in reordering their concepts of the pattern of change. They are also faced in the U.S.A. with a resurgence of Creationist concepts (see Shea (1983) for a discussion). However, it seems fair to characterize Creationist ideas as a political irritant for the scientific community, and one that has failed to muster any momentum intellectually.

Ergodicity

Physicists were confronted with a major observational problem when originally investigating molecular behavior, namely that molecules move very rapidly in comparison to the duration of observation. They sought to overcome this difficulty by using a concept that is fundamental to statistical mechanics – the ergodic theorem or hypothesis (from the Greek *ergon*, meaning work or energy, and *hodos* meaning road; Paine 1985, p. 2). Boltzmann is generally credited with the initial formulation of the idea, as well

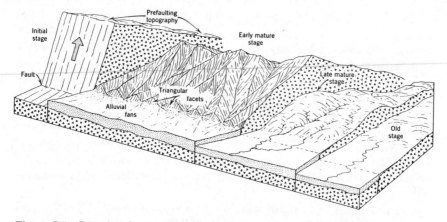

Figure 5.1 Postulated evolution of a fault scarp through time. Terminology and forms are strictly Davisian, but the use of a spatial surrogate to create such a temporal sequence is widely used in geomorphology. (From Strahler (1971).)

as coining the term, but some aspects of it (not termed ergodic at the time) may be traced all the way back to 1713 and Bernoulli's (weak) law of large numbers (Brown 1976).

In simple terms the gist of the ergodic hypothesis is that the mean of observations of an individual made over time is equal to the mean of observations made of many individuals at a single moment in time over an area (see Thornes & Brunsden (1977, p. 24) for a nice analogy). In rather more formal terms the concept is that, with respect to molecular behavior, phase and time averages are assumed to coincide on each surface of constant energy (Parry 1981, p. 1). Among the most important attributes of formally defined ergodicity are that: (1) it is founded on rigorous statistical concepts; (2) it is mean values of spatial and temporal distributions that are central to the concept; and (3) time is used as a surrogate measure for space.

While ergodicity has been examined in great detail by mathematicians, its formulation in geomorphology is very crude. Chorley & Kennedy (1971, p. 349) stated simply that it is "the hypothesis that, under certain circumstances, space and time can be considered as interchangeable". While they noted its rigorous foundations, they did not discuss them; nor has anyone else in geomorphology prior to Paine (1985). As an initial step the manner in which "ergodicity" has usually been presented in geomorphology will be examined.

Consider the scarp shown schematically in Figure 5.1. The diagram purports to represent evolution of a single scarp through time. Clearly, no one has ever seen such a sequence completed. The composite was created by field examination of existing scarps (a spatial distribution) and an "ergodic" assumption; namely, that this spatial distribution is a surrogate measure of the relevant temporal one.

Several points emerge when comparing geomorphologists' and physicists' use of ergodic concepts. Geomorphologists have made little, if any, attempt to study the concept quantitatively. They have been much more concerned with substituting space for time, whereas physicists originally pursued exactly the opposite goal. Originally ergodicity embraced the idea that "time averages can be replaced by space averages" (Craig 1982b, p. 83), but the "average" component has not been pursued much by geomorphologists. Lastly, geomorphologists have generally tended, as in Figure 5.1, to translate spatial distributions into temporal sequences; the incorporation of a specific order (sequence) in the temporal series is not a component of ergodicity and represents a profound and questionable shift (Thornes & Brunsden 1977, p. 177).

If ergodicity is used rigorously, the temporal and spatial distributions have the same attributes; for example, if 15% of the population is in a particular state at any moment in time, it is appropriate to assume that each individual in the population spends 15% of its time in that same state. This kind of precision has not been applied in geomorphology, but without it space–time or time–space substitution is not truly ergodic (Paine 1985, p. 4). Paine suggested that an appropriate term for what is normally undertaken in geomorphology, substitution of space for time without statistical rigor, is "location-for-time substitition". In suggesting this, he rejected Thornes & Brunsden's (1977, p. 25) proposed term "space–time analog" because he felt it failed to differentiate sufficiently between true ergodicity and geomorphic practice.

Paine went on to discuss the use of location-for-time substitution and formal ergodicity in geomorphology. He suggested that geomorphologists are generally interested in modeling either equilibrium forms (also called characteristic forms by Brunsden & Thornes (1979)), where the landform is assumed to be adjusted to the processes acting upon it, or relaxation forms, where a landform is assumed to change its form through time in response to a change in inputs or conditions.

When studying characteristic forms it is necessary to make several important assumptions (Paine 1985, pp. 6–9), including that: (1) process controls and system structure are invariant over time; (2) sources of variation are understood; and (3) characteristic form conditions are actually present today. Quite obviously this is a very demanding set of constraints.

Modeling relaxation forms, or tracing the sequence of forms during landform evolution, has always been focal to geomorphology. However, creation of such models requires many assumptions. Foremost among them is the ability to place the array of landforms into what is assumed to be the correct temporal sequence – this must be done on some theoretical basis. As a result, the confidence with which this initial ordering task may be undertaken becomes the limiting factor. In general terms it may be seen as being achieved at a nominal, ordinal, interval, or ratio scale, with the usual increase in desirability at the higher scales (Paine 1985). Nevertheless, at any scale, time is the issue, and any other source of variation represents contamination. Establishment of purely temporally based variability is, of course, extremely difficult in geomorphology.

There are only one or two examples of truly ergodic modeling in geomorphology. Melton's (1958b) attempt to derive growth laws for drainage basins using only mature drainage basins is an early example incorporating an equilibrium model. It should be noted that Melton (1958b, p. 49) himself described his use of the ergodic concept as heuristic and not rigorous. Carter & Chorley (1961) also examined drainage-basin development using ergodicity, but they incorporated a relaxation model. They used drainage-basin order (a spatial measure) and transformed it into a temporal measure. Their argument was based on the assumption that all basins start as first-order basins and, therefore, basin order is actually a measure of basin age.

Ergodicity and location-for-time substitution is actually best seen in geomorphology in a number of studies of scarp evolution. This trend was initiated by a pioneering study by Savigear (1952). He examined a sequence of slope profiles along part of the coast of South Wales where the eastward growth of marine deposition had isolated slopes from marine erosion at different times. Therefore, slope profiles at different locations behind the depositional barrier represented a sequence of slopes that had experienced differing periods of post-marine, sub-aerial modification. Brunsden & Kesel (1973) undertook a similar study along a river bluff sequence of known and differing elapsed times since the termination of fluvial undercutting. While such studies have been cited as ergodic, they are not truly so (duly noted by Brunsden & Kesel (1973, p. 576)) because the temporal sequence is actually known to exist and is not postulated from theory. Carson (1971) and Nash (1984) represent scarp development modeling in which ergodicity is more faithfully represented. Carson provided a quasi-quantitative estimate of the amount of time each slope profile spends in particular configurations, deriving his estimates from spatial sampling. Nash synthesized a theoretical model of scarp evolution through time and then integrated his theoretical, temporal model with spatial sampling of contemporary scarps to estimate their age.

One way in which change over time has been studied in biology is allometry. Allometry is "the study of proportional changes correlated with variation in size of either the total organism or the part under consideration. The variates may be morphological, physiological, or chemical" (Gould 1966, p. 629). Bull (1975a, 1977) has promoted the use of allometry in geomorphology, while papers by Cox (1977) and Church & Mark (1980) highlighted some of the critical issues involved in its use. As Paine (1985, p. 6) observed, allometry may be viewed as a special case within ergodicity.

There are two kinds of allometry: dynamic allometry describes changes in relationships between parts of a single individual through time; static allometry describes the same kinds of variation in an entire group of varying sizes at any moment in time. In practice, both kinds of relationships are often expressed as power functions. A frequent result of comparing dynamic and static allometric studies of the same phenomena is that the relationships differ. This outcome may be compared to an established geomorphic relationship; it has long been known that even within a single river relationships between water discharge and sediment discharge differ when the results from a single station sampled at various times are compared to results from several different stations all sampled at the same time. If equations for dynamic and static allometric studies of the same features differ at a statistically significant level (still uncertain in most instances), this would negate ergodicity. Church & Mark (1980) provided an excellent review of allometry, its use, and limitations within geomorphology.

The utility of ergodicity in geomorphology is really still unknown. This stems from the simple fact that it has never been formulated in anything like formal terms within the discipline. Craig (1982b) gave one of the most elaborate presentations in geomorphology incorporating ergodic principles and concluded that it was useful; however, Church (1983) noted some limitations in Craig's model. If geomorphologists have not really employed ergodicity, it seems that they should. Furthermore, if they think that they have (whereas, in reality, they have not), there is an unrecognized crisis that needs confronting: "What theoretical underpinning is there, if any, for the widespread assumption that spatial distributions of landforms reveal anything about the temporal evolution of individual landforms?"

Conclusions

Rejection of uniformitarianism seems appropriate for geomorphologists, but it should be made with an appreciation of what it achieved as well as of why it must be replaced. There would appear to be an inherent tension between magnitude and frequency concepts (a modern derivative of uniformitarianism) and neocatastrophism (the modern equivalent of catastrophism), one that is heightened as the period of interest lengthens. Nevertheless, it is necessary to appreciate the strengths of both approaches and seek to weld them into a viable combination.

In examining formal ergodicity it is apparent that geomorphologists really have no substantive grasp of this issue. Nevertheless, it would be absurd to believe that there is no relationship whatsoever between today's landscape and the evolution of individual landforms through time. The problem is that more effort needs to be expended on constructing and considering the formal relationships between spatial and temporal distributions that geomorphologists are going to espouse. There must be some formally and theoretically defined rationale for believing that present landscapes and landforms reveal something about their paleo-equivalents.

In short, uniformitarianism and ergodicity strike at the very heart of geomorphology. Nothing is achieved by criticizing them negatively or rejecting them, unless it results in replacing them with something that appears more satisfactory. For without these theoretical constructs or viable alternatives to them there can be no discipline of geomorphology.

6 Time in geomorphology

The nature of geomorphic time

It is all but impossible to exclude time from any discipline. However, in geomorphology its role is pivotal because there has been a traditional preoccupation with landscape change over time. Time would certainly have been less important in geomorphology if geomorphologists had originally focused their attention on just the mechanics, chemistry, and physics of change. Despite the importance of time, there are no fundamental or standard time periods for geomorphic study. As there is a general, positive correlation between the size of a landscape unit and the time that it takes for a unit to change, the time periods embraced by geomorphic studies may vary with the subdisciplinary approaches outlined in Chapter 3.

Geomorphologists have usually considered landscape behavior over time in one of two ways (see, e.g., Brunsden & Thornes 1979, Paine 1985). One way has been to examine landscape change as a response to some sort of input. Following Brunsden & Thornes, this is called a relaxation-time model. The second approach has been to model landscapes or landscape units as if they were in balance with the inputs acting upon them. Brunsden & Thornes (1979, p. 464) dubbed these characteristic-form models. It should be noted that such models are often called equilibrium models; however, equilibrium is a complex issue and use of it in this context invariably implies more than can be truly verified.

Relaxation-time models dominated geomorphology for decades, if only because of the pre-eminence of William Morris Davis's ideas in the English-speaking world. While Davisian geomorphology will be considered at length in Chapter 10, some of its characteristics must be briefly reviewed here because, while not the first geomorphologist to consider landform evolution (Chorley et al. 1964), Davis was certainly one of the most influential.

The Davisian model of landscape change implicitly contained many attributes of a closed system (Chorley 1962, p. B2) and may be called a time-decay model (Thornes & Brunsden 1977, pp. 19–20). In it uplift occurred, thereby providing energy, and subsequently the landscape evolved at an ever-slower rate over the period of an "erosion cycle" as the energy was used. During the course of an erosion cycle landscape morphology changed; hence its designation as a relaxation-time model.

Not only was the Davisian model focused upon change over time but, subsequent to uplift, change followed a gradualist pattern. Stoddart (1966) noted the influence of Darwin's ideas upon Davis. Davis was preoccupied with change over time, the Darwinian concept of evolution. However, Darwin's great contribution was not to suggest change, but rather to suggest that randomness was the explanatory factor. Nevertheless, he vacillated very greatly through the years on the critical issue of randomness (Stoddart 1966, p. 696). In embracing Darwin's ideas, Davis focused upon evolution, but in his version of evolution landscape change was not only gradualist, it was also inevitable. Consequently, the random component was lost and as Stoddart (1966, p. 688) noted so elegantly " . . . what for Darwin was a process became for Davis and others history."

Thus, for many years geomorphologists were occupied with the timespans of Davisian erosion cycles during which landscapes evolved or changed morphology. The duration of erosion cycles was considered to be long (many millions of years), but definitely shorter than geologic time. Lacking the means to measure time at appropriate timescales (obviously isotope techniques had not been developed), time was treated rather passively – conceptually, it began at uplift and then merely elapsed.

While it is easy to understand why time itself was treated so passively in Davisian geomorphology, despite the great emphasis on stage, treatment of time by Davisian geomorphologists became something of an Achilles heel. The enormous, but sketchily treated, timespans invoked in Davisian geomorphology became more manageable as dating techniques emerged, and with this came new problems as hitherto unappreciated conflicts emerged.

Radical departure from Davisian constraints was represented by Strahler's (1950) paper in which he emphasized the need to examine landscapes, not from the Davisian perspective in which available energy is always decreasing, but from one in which there is constant change in available energy. Subsequently, Chorley (e.g., Chorley 1962) has done much to promote this viewpoint. The core of the issue is abandonment of a closed-systems approach and acceptance of an open-systems one. The most critical difference to emerge from this shift is that in an open system landforms may attain equilibrium. Clearly then this forms the foundation for characteristic-form models in which it is assumed that there is some sort of balance between inputs and landscape morphology.

With the introduction of equilibrium concepts, a long tradition of relaxation-time modeling, and some satisfactory techniques for measuring time, geomorphologists found themselves with some important choices to make and some means of evaluating them. In short, time itself became a lively issue rather than just a passive commodity. Wolman & Miller (1960) introduced the concepts of magnitude and frequency, while Schumm & Lichty (1965) evaluated the role of time itself. Nevertheless, it was not until 1977 that a textbook, *Geomorphology and time*, focusing on the role of time in geomorphology, emerged (Thornes & Brunsden 1977).

From the web of ideas that constitutes the way geomorphologists incorporate time in their work, a number of important themes may be selected for separate treatment: (1) classification and sampling of time; (2) the concept of frequency (and magnitude); (3) landform response to inputs; and (4) equilibrium.

Treatment and sampling of time

Schumm & Lichty (1965) initiated a quantitative evaluation of the role of time in geomorphology by subdividing it and showing the significance of the subdivision. The quintessence of their perspective appeared in the first sentence of their abstract (the perfect location by the way!) (Schumm & Lichty 1965, p. 110):

> The distinction between cause and effect in the development of landforms is a function of time and space (area) because the factors that determine the character of landforms can be either dependent or independent variables as the limits of time and space change.

While Schumm & Lichty's categories are only one of many possibilities, geomorphologists could no longer afford to view time elapsing in a monotonic fashion, but were forced to treat it as a living issue that in its varying duration fundamentally altered the matters under consideration.

Schumm & Lichty divided time into three categories, those of cyclic, graded, and steady time. Each was described and characterized by them and may be summarized as follows.

(1) *Cyclic* – The timespan of a Davisian erosion cycle. During such time a fluvial system may be thought of as an open system with no specific or constant relationships between independent and dependent variables. Time itself is the most important independent variable and an entire basin or any component may be studied.

(2) *Graded* – A short segment of cyclic time during which a graded condition or dynamic equilibrium exists. The situation is dominated by negative feedback (self-regulation) and there are either a series of fluctuations about or approaches to a steady state. Graded time cannot occur throughout the system because relief reduction is ongoing, but components of the system may experience graded time.

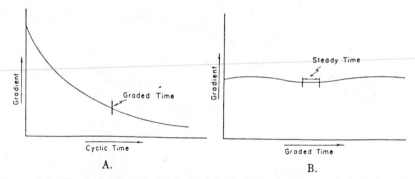

Figure 6.1 The behavior of channel gradient during cyclic, graded, and steady time. (a) Progressive reduction of channel gradient during cyclic time. During graded time, a small fraction of cyclic time, the gradient remains relatively constant. (b) Fluctuations of gradient above and below a mean during graded time. Gradient is constant during the brief span of steady time. (From Schumm & Lichty (1965).)

(3) *Steady* – A brief period during which some portions of the system may be truly time-independent, that is, unchanging through the time period involved.

The nature of landscape change during the three types of period is shown schematically in Figure 6.1. Schumm & Lichty did not offer actual limits in years for the three categories and precise delimitation is infeasible. However, Thornes & Brunsden (1977, p. 15) suggested reasonable approximations: cyclic, 10^6 y; graded, 10^2 y; steady 10^{-2} y. The probable interplay between variables at each timescale was summarized by Schumm & Lichty (1965) in tabular form and is reproduced here as Tables 6.1 and 6.2.

Clearly, the timespan addressed by a geomorphologist is intimately related to his/her objective, but it is also greatly constrained by a combination of available techniques and research time. Lewin (1980, pp. 5–6) pointed to four broad-scale categories in which data may be generated. First, there is direct field observation and/or measurement, generally involving timespans of less than 10 years. Second, there are historical data, although these were invariably not created for the purpose at hand. Third, there are opportunities to date various kinds of surfaces (e.g., soil development). Fourth and finally, geomorphologists may turn to purely stratigraphic methods, incorporating both relative and absolute dating techniques.

Time has many important attributes when it is examined as a geomorphic variable. Perhaps the most important of these is the inherent characteristic of direction: on a macroscale, time is irreversible; and, at any scale, events are locked in place in a series. These are fundamental qualities for geomorphologists who are primarily interested in rates of operation, the

Table 6.1 The status of drainage-basin variables during timespans of decreasing duration. (From Schumm & Lichty (1965, p. 112).) Variables are organized from top to bottom in presumed order of increasing dependence.

Drainage-basin variables	Status of variables during designated timespans		
	Cyclic	Graded	Steady
(1) Time	independent	not relevant	not relevant
(2) Initial relief	independent	not relevant	not relevant
(3) Geology (lithology, structure)	independent	independent	independent
(4) Climate	independent	independent	independent
(5) Vegetation (type and density)	dependent	independent	independent
(6) Relief or volume of system above base level	dependent	independent	independent
(7) Hydrology (runoff and sediment yield per unit area within system)	dependent	independent	independent
(8) Drainage network morphology	dependent	dependent	independent
(9) Hillslope morphology	dependent	dependent	independent
(10) Hydrology (discharge of water and sediment from system)	dependent	dependent	dependent

direction and duration of events, and the memory and relaxation times of landforms. In process studies, velocity and acceleration also loom as fundamental temporal issues. All of these attributes were identified by Thornes & Brunsden (1977, pp. 2–3), who went on to outline some basic approaches to viewing the pattern of events and also the sampling of time.

Events may be regarded as isolated if they have truly "one time only" frequency, but most are much more appropriately viewed as having continuity, fluctuation, and sequence (Thornes & Brunsden 1977, p. 3). Furthermore, their dynamic behavior may be random, unidirectional or cyclic. Finally, any event may be viewed as reversible or irreversible.

Thornes & Brunsden (1977, pp. 5–9) also summarized the manner in which time is sampled by geomorphologists. Observations and measurements may be drawn from continuous, quantized, discrete, or sampled time; each may be characterized as follows.

(1) *Continuous* – Observation is unceasing. However, note that continuous observations are virtually never analyzed without first being reduced in some fashion.

(2) *Quantized* – Division of continuous time into segments deemed useful or at least familiar (e.g., hour, day, month, year). The manner of summing the information for the period of interest (e.g., is a daily value derived from the mean of values taken every minute,

Table 6.2 The status of river variables during timespans of decreasing duration. (From Schumm & Lichty (1965, p. 116).) Variables are organized from top to bottom in presumed order of increasing dependence.

River variables	Status of variables during designated timespans		
	Geologic	Modern	Present
(1) Time	independent	not relevant	not relevant
(2) Geology (lithology, structure)	independent	independent	independent
(3) Climate	independent	independent	independent
(4) Vegetation (type and density)	dependent	independent	independent
(5) Relief	dependent	independent	independent
(6) Paleohydrology (long-term discharge of water and sediment)	dependent	independent	independent
(7) Valley dimension (width, depth, and slope)	dependent	independent	independent
(8) Mean discharge of water and sediment	indeterminate	independent	independent
(9) Channel morphology (width, depth, slope, shape, and pattern)	indeterminate	dependent	independent
(10) Observed discharge of water and sediment	indeterminate	indeterminate	dependent
(11) Observed flow characteristics (depth, velocity, turbulence, etc.)	indeterminate	indeterminate	dependent

every hour, or just from daily maximum plus daily minimum divided by 2?) and the duration of the period selected (day or month, etc.) will determine the loss of information.

(3) *Discrete* – Focus is placed upon the length of the time increment selected, e.g., per day, per month, per year. It is a device often associated with frequency.

(4) *Sampled* – The process of interest is recognized as continuous, but sampled for short periods (i.e., quantized). Note the flexibility inbuilt into sampling something one day a month; it is possible to express the data as daily values (if measurements are taken for 24 h), or to take the total difference that occurs between two sampling days and express it as a monthly value.

Frequency

One of the most widespread ways in which geomorphic inputs have been measured temporally is as frequencies. Commonly, the occurrence of an event is not monitored alone but is combined with estimation of the event's magnitude (e.g., Gardner 1980). If both attributes are measured it becomes

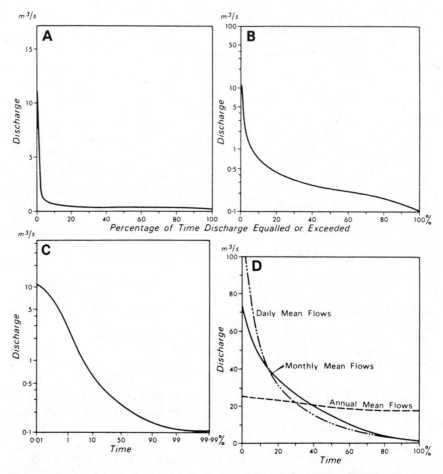

Figure 6.2 A variety of ways of displaying river discharge (magnitude) against frequency. (a) x and y coordinates are *both* linear. The main advantage is that the area beneath the curve is directly proportional to the total discharge. The main disadvantage is the loss of detail concerning low flows. (b) Discharge plotted on a logarithmic scale to overcome the difficulty in (a). (c) Normal probability scale is used on the duration ordinate to highlight extreme values. (d) An illustration of how the shape of the curve, especially its slope, is influenced by the time increment used, in this case daily, monthly, and annual mean flows. (From Gregory & Walling (1973).)

possible to plot cumulative frequency (Fig. 6.2); such a plot has the advantage of enabling the researcher to make statements about the frequency or proportion of time equalled or exceeded by events of known magnitude.

When many natural phenomena are plotted it is usually found that zero values are uncommon, that the frequency of occurrence increases to some moderate magnitude, and that when very large magnitudes are examined frequency once again drops off. Even more specifically, it is found that many natural phenomena exhibit a log normal distribution (e.g., Wolman & Miller 1960, Caine 1968). In these, and other nonnormal, distributions data are best transformed prior to analysis (see Gregory (1978, pp. 53–9) for simple illustrations).

Wolman & Miller (1960) recognized the general relationship between magnitude and frequency (i.e., moderate events are commonplace and large events rare) and sought to quantify the answer to the question: "Are small or large events the most important in geomorphology?" They used fluvial geomorphology as their primary research vehicle and using the notion (Wolman & Miller 1960, p. 56) that

$$\frac{(\text{sediment carried by given flow}) \times (\text{frequency of given flow})}{\text{total sediment transported}} \times 100$$

= percentage of total sediment carried by flows of different magnitude

they determined that at least 50% of sediment was carried by flows of a magnitude that occurred at least once every year. In related calculations they illustrated the general principle that events of intermediate magnitude, and therefore also of intermediate frequency, are generally dominant in fluvial sediment transport. This paper is a benchmark, initiating an appreciation of the significance of magnitude versus frequency in determining quantitatively the relative importance of individual events.

An allied concept is that of recurrence interval (*RI*). By plotting extreme events (usually the largest event of each year) on specially designed graph paper (usually called "Gumbel paper") it is possible to calculate the frequency of an event of specific magnitude and the probability of an event of specified magnitude occurring in any year, the latter being the inverse of its frequency. Such calculations are widely used in meteorology and hydrology, and are summarized and illustrated in Dunne & Leopold (1978, pp. 51–66, 80–1). There are distinct limitations to these calculations, particularly with short records and large events having an *RI* approaching the length of the record; nevertheless, they provide a much needed and useful means of extrapolation. When evaluating meteorological and hydrological events for their geomorphic significance, even more uncertainties are introduced because of other controls (e.g., soil and rock characteristics).

One basic problem with frequency and magnitude concepts is that they are founded on the concept of "stationarity"; this term simply means that the pattern of events has a mean and variance that do not vary through time. Stationarity is essentially equivalent to substantive uniformitarianism. As a result, "the frequency distribution of events does not sufficiently describe the properties of a sequence" (Church 1980, p. 14). Three primary problems emerge, those of trend, persistence, and intermittency. Trend is generally well understood, but it is important to appreciate that trends include cyclic patterns. Persistence occurs when adjoining values in a series limit one another; this means that the true variability of the system may not emerge in a short record. Such behavior may be investigated using Markovian analysis (Collins 1975) and other models designed to assess autocorrelation (e.g., Box & Jenkins 1976).

Intermittency is the least well-known of the three problems and is characterized by nonperiodic clustering of similar values over long periods. Such clustering may produce cumulative effects that have fundamental geomorphic impact. Church (1980, pp. 15–18) discussed analysis of data sets exhibiting this type of behavior (so-called "Hurst phenomenon") using a technique called the rescaled range (Hurst 1951).

Concern with frequency and magnitude may be legitimately taken to extremes in geomorphology (although not as far as in geology). Given the great duration of the landscape record, it is worth examining the importance of events of very low frequency and very great magnitude. A perspective that invokes the importance of such events is normally called neocatastrophism. It is a topic already considered in Chapter 5 from a nonquantitative and historical perspective. Here we will look at it briefly, but from a statistical viewpoint.

Gretener (1967) provided a good demonstration of the statistical probabilities that emerge when rare events (where rare is usually measured on a human timescale) are examined on a geologic timescale. He provided numerous illustrations, but one will suffice here: "an event with a yearly probability of 10^{-5} is 'bound to' happen (95% probability) at least once in half a million years" (Gretener 1967, pp. 2199–200). This frequency is clearly likely to be of significance in many geomorphic contexts and Dury (1980) provided an interesting array of geomorphic calculations; the emptying of glacial Lake Missoula and the creation of the Channeled Scablands of eastern Washington State once again serve as a prime example.

As ideas involving magnitude and frequency have evolved, important additional concepts have emerged. Wolman & Miller (1960) originally considered work as manifested in fluvial sediment transport. However, this approach did not address landform initiation or creation directly (presumably it does address landforms indirectly as there must be a sediment source). There is a fundamental distinction between events that transport sediment in a stream channel (virtually all events do this to some degree) and those events that actually create a new landform (i.e., initiate a gully or create a new stream course). Furthermore, it is unlikely that events that transport the maximum sediment load will have the same magnitude and frequency as events initiating new landforms. Wolman & Gerson (1978) addressed this latter issue and introduced the important modifying concepts of relative timing and relative magnitude (Wolman & Gerson 1978, p. 190), leading to an examination of effectiveness and recovery times.

They pointed out that the geomorphic significance of a formative event is not its absolute magnitude and frequency, but rather its frequency with respect to local recovery (healing) times and its sequential position. For example, if two 50 y events take place in consecutive years, they would be of approximately the same magnitude, but the second one should clearly achieve greater geomorphic results because it is unlikely that the landscape would have fully recovered from the event of the preceding year. Wolman & Gerson examined the problem at length and calculated that in tropical areas formative events may occur every 1–2 y and as infrequently as once every 100 y or more in temperate regions.

One possible criticism of Wolman & Gerson's work is that they couched the concept of recovery time exclusively in terms of revegetation. This may well be unavoidable, but revegetation without obliteration of the change in landform may not mean that the area is once again offering the surface resistance that it originally had. Subsequently, Brunsden & Thornes (1979) attempted to quantify the relationship between events (representing the tendency for change) and surface resistance (representing the resistance to change).

Brunsden & Thornes (1979) suggested dividing geomorphic time into the time it takes a landform to adjust to a new form (relaxation time) and the period over which the form will then persist as it is able to absorb the inputs acting upon it (characteristic-form time). Using these ideas it is possible to assess the sensitivity of landscape or landscape elements to change in terms of a transient-form ratio TF_r, which equals mean relaxation time divided by mean recurrence time of events (that produce change).

A significant portion of modern geomorphology evolved with the implicit assumption that fluvial systems and interfluves are fully integrated. A number of researchers (e.g., Caine 1974) have suggested that, in fact, linkage between the two may be weak, selective, or infrequent. This has led to recognition of the fact that magnitude and frequency relationships must be considered separately for differing sections of the landscape. This trend is now firmly established and is exemplified in Wolman & Gerson (1978), Brunsden & Thornes (1979), Dury (1980), and Starkel (1976). It is important to appreciate that the need for spatial stratification also occurs on a larger scale; magnitude and frequency concepts were originally developed using data from the humid temperate regions. Initially, values from this work were assumed to be generally valid, but it is now widely recognized that they will vary regionally (e.g., Selby 1976, Wolman & Gerson 1978).

The need to stratify a landscape to evaluate its magnitude–frequency regime quickly leads us into spatial issues – the subject of the next chapter.

However, one other issue that stems from magnitude–frequency concepts must also be briefly mentioned here. Magnitude and frequency must be calculated for every process individually and with the great variety of geomorphic processes it quickly becomes extremely difficult to compare them; for example, how should one compare the relative importance of rockfall and soil creep to landscape development? Reduction of all individual processes to a uniform measure is often an excellent means of permitting valid comparison of diverse processes. Rapp (1960) undertook this step by expressing all processes in terms of the number of tonnes of material moved vertically downslope; Caine (1976) followed a similar scheme, but expressed everything in energy units. In either instance, comparison is greatly enhanced by use of a single metric.

Magnitude and frequency provide an important perspective from which to view geomorphology. Elaboration of the original idea has not negated the underlying principles; rather, it points to an acceptance of them. Our grasp of magnitude and frequency has now matured and is more sophisticated; inevitably this includes recognition of some limitations and complexities. Even inputs themselves may be stratified according to their nature. Brunsden & Thornes (1979, pp. 468–9) suggested recognition of pulsed and ramp inputs. The former is an input of short duration in comparison to the time period under consideration; the latter type of input is sustained at the new level once initiated. Regardless of input type, several different responses are possible and these will now be examined.

Change through time

The response of a landform to either a pulsed or ramp input may be to absorb it without change, to adjust to it without perceptible change, or to change – often over a longer period than the duration of the input itself. If the input is absorbed without change, the landscape may be exhibiting some sort of characteristic form or equilibrium behavior. This is a sufficiently complex topic that it will be treated separately in the next section. If change occurs, a number of interesting concepts may be relevant to describe landform behavior.

One important, but often imperceptible, change in response to a pulsed input is hysteresis. This concept is one of "incomplete reversibility" (Williams 1982, p. 53) and occurs in conditions as diverse as soil wetting and drying (Fig. 6.3a), and water discharge–sediment discharge relationships during floods (Fig. 6.3b). In simple terms the idea is that, when something moves from condition A to condition B and then back to A, the pathways from A to B and then back from B to A differ. Furthermore, condition or position A may not actually be regained but A^1 (i.e., something measurably different from A) may be the final state. When considering something like

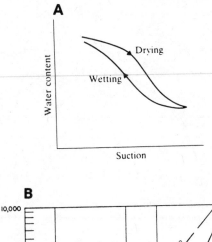

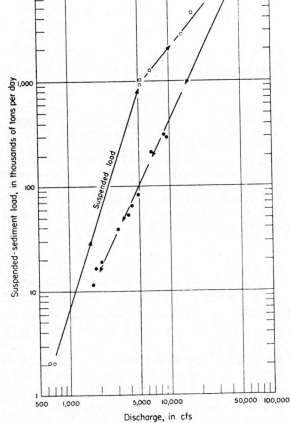

Figure 6.3 (a) Hysteresis in suction–water content relationship. (b) Hysteresis between suspended sediment concentration and discharge during the rising and falling limbs of a flood on the San Juan River. (Part (a) from Williams (1982); part (b) from Leopold & Maddock (1953).)

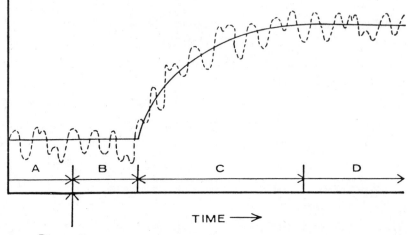

Y

A   B   C   D

TIME ⟶

Disruption

Figure 6.4 Graphic representation of the response of a geomorphic system subjected to disruption. The full line indicates mean condition, and the broken line represents actual values. A system parameter Y (some dimensional or spatial characteristic of the system such as length or width) is in a steady state during period A. After the climatic or human disruption, new conditions are internalized by the system during period B, the reaction time. During the relaxation time, period C, the system adjusts to the new conditions. The rate law provides a model for the system change during this period. A new steady state is established in period D, resulting in new dimensional characteristics as signified by a new mean value for Y.

Not all systems experience steady states. In some cases the initial condition might be in dynamic equilibrium and be drawn as a sloping line in times A and B. After a threshold is passed, system adjustments in period C would be the same as shown in the diagram. The new conditions in time D might also be represented by a sloping line indicating dynamic equilibrium. (From Graf (1977).)

the loading and unloading of clays, hysteresis implies that the system exhibits memory and behavior will depend to some degree on the unique loading–unloading (i.e., stress) history of the material.

In the case of long-term response to disruptive input, Graf (1977) found that the rate law used by chemists and physicists could be applied successfully in at least one instance. Such an approach identifies reaction time (the time it takes the system to absorb the disruption) and relaxation time (the time during which the system adjusts to the new conditions) (Fig. 6.4).

Studies such as Graf's have not been undertaken widely, although Allen (1974) and Knox (1972) have discussed the concepts with respect to valley alluviation and sedimentation, respectively. Indeed the rate law, most familiar to us as the basis for half-life statements concerning radioactive decay, may not be valid in every circumstance or even in most instances. However, whatever the detailed circumstances surrounding a particular situation it is critical to appreciate that a landscape or individual landform is not necessarily going to respond to an input immediately, nor in the case of a pulsed input is the response necessarily going to cease once the input does.

The schematic diagram in Figure 6.4 reveals nothing about the particular form changes that will occur during the relaxation time. Such a sequence may be quite complicated; indeed Schumm (1979) has shown that in the case of river channels responses may be very complex. His results will be illustrated in the following section on equilibrium.

Entropy and equilibrium

Chorley & Kennedy (1971, p. 348) define equilibrium as "a condition in which some kind of balance is maintained". Traditionally, equilibrium has played an important role in geomorphology. Given that landforms are considered to change over time, it follows that geomorphologists will be particularly interested in conceptual frameworks that predict and/or retrodict the pattern of change. One set of concepts that goes some distance to fulfilling this task is that of various types of equilibria. Equilibrium is a complex concept that does not exhibit a single pattern. Nor does equilibrium necessarily occur with respect to all measures simultaneously. However, one context in which energy equilibrium has been defined precisely is the concept of entropy.

Entropy is formally defined as the free energy in an isolated system. Free energy is that energy available for conversion to work (and for a geomorphologist one byproduct of work is landform change), and an isolated system is one into which, by definition, no energy or mass may be imported or exported. Energy can only move down an energy gradient. Therefore, if energy in an isolated system is unevenly distributed at the time of initiation it will move down available gradients. The result of this transfer is that energy within the system becomes more evenly distributed, gradients become lower, and while the process continues it does so at an ever-decreasing rate. The ultimate result is that energy becomes evenly distributed and, regardless of whether the absolute total of energy within the system is high or low, there will be no free energy when the distribution of it is even and the gradients eliminated. Entropy is said to be at a maximum when there is no free energy in a system.

An analogy may serve to reinforce the concept of entropy; Davidson (1978, p. 59) cited a nice example presented by Reynolds (1974). Reynolds considered a tray with 100 red jumping beans and 100 white jumping beans on it. There is only one way for the tray to be all red on one side and all white on the other; this is a perfectly ordered situation and the entropy is zero. As the beans begin to jump around, the tray turns pink, and there are an infinite number of internal distributions that could produce this pink state. Clearly, the pink state is a statistical concept, and as the randomness or disorder increases on the tray, so does the entropy. When entropy is at a maximum there is no free energy and, therefore, the system is in equilibrium.

The concept of entropy was developed in thermodynamics and statistical mechanics. In this context, as energy distributions tend from the improbable to the probable, entropy increases and free energy decreases. However, as Leopold & Langbein (1962, p. A2) noted, entropy has been used successfully in other contexts where it has become essentially a synonym for probability. These authors pointed out there that:

> ... the entropy of a system is a function of the distribution or availability of energy within the system, and not a function of the total energy within the system. Thus, entropy has come to concern order and disorder.

Leopold & Langbein went on to note that, when entropy is applied to issues outside thermodynamics, it is necessary to identify the alternative system. They then demonstrated how entropy may be used to explain the widely recognized principle of least work. Among the important geomorphic systems thought to exhibit least-work behavior are the longitudinal profiles of rivers, hydraulic geometry, and drainage networks.

In treating entropy statistically, Leopold & Langbein (1962, pp. A3–4) showed that it may be viewed as the logarithm of probability and that it is maximized when all the various states have equal probability. However, they made one other very interesting observation and that is that the maximum value of entropy is very insensitive and large deviations from the absolute maximum may well occur. The behavior of the mathematical function describing entropy suggests that there will be rapid movement away from highly abnormal distributions (or configurations), but considerable latitude in the central zone. Of course this fits very well with a decay function that implies that rapid change will occur at first, followed by a decreasing rate of change over time.

In geomorphology any distribution embracing random components can only be developed within constraints imposed by a variety of "ordering" controls, such as lithologic variation, and entropy is no exception to this principle. By definition, once created, entropy cannot be destroyed. However, entropy is an equilibrium concept associated with isolated systems and in open systems other types of equilibrium may be expected.

Figure 6.5 shows two equilibria associated with closed systems (static and thermodynamic) as well as several others that depend upon the existence of systems permitting energy to be transported across their boundaries. Chorley & Kennedy (1971, pp. 201–9) defined each equilibrium type as follows.

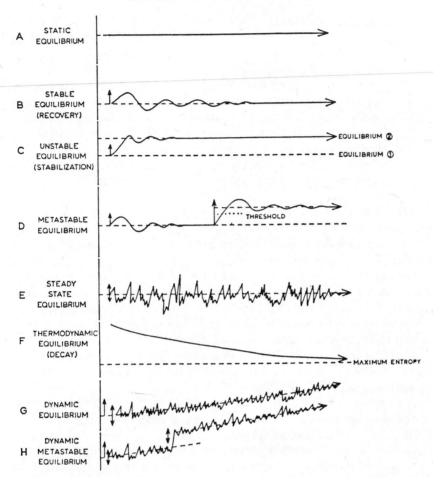

Figure 6.5 Diagrammatic illustration of eight equilibrium conditions. (From Chorley & Kennedy (1971).)

(1) *Static* – A situation in which a balance occurs such that some system properties are static (unchanging), absolutely and relatively, over a period of time.

(2) *Stable* – A tendency to revert to a previous condition after experiencing limited disturbance.

(3) *Unstable* – A situation where a small disturbance results in continued movement away from the old equilibrium, usually terminated upon arrival at a new stable equilibrium.

(4) *Metastable* – Stable equilibrium exists until some incremental change (a trigger mechanism) pushes the system across a threshold into a new equilibrium.

(5) *Steady state* – No trend in mean value over the period of interest but

numerous small-scale oscillations about the mean. [Note this is "stable" with many disturbances.]

(6) *Thermodynamic* – A tendency toward maximum entropy as implied by the second law of thermodynamics.

(7) *Dynamic* – Balanced fluctuations about a mean value that itself has a trending, nonrepetitive, mean value. A tendency toward a steady state with a trending mean commonly means that equilibrium is rarely, if ever, attained and is often labeled "quasi-equilibrium".

(8) *Dynamic metastable* – A combination of dynamic and metastable equilibria, whereby minor fluctuations about a trending mean value are interspersed by large jumps as thresholds are crossed.

A number of these equilibria concepts have played important roles in the way geomorphologists have viewed landscape evolution. However, they also embrace some supporting concepts that must be briefly outlined, prior to discussing the equilibria themselves.

Movement away from and return to some mean value requires a mechanism. Most commonly movement that continues away from the mean after a disturbance is called positive feedback. The opposing tendency whereby a disturbance is dampened or obliterated and the trend is for a return to the original values is called negative feedback. Both of these mechanisms are integral to systems modeling of geomorphology (promoted by Chorley; see Chorley & Kennedy (1971) for a good treatment) and will be discussed more fully in Chapter 12.

The notion that a small or incremental change may provide a large response is founded on threshold concepts. Schumm (see, e.g., Schumm (1979) for a good summary) has recognized extrinsic and intrinsic thresholds in geomorphology. A threshold itself is a condition that marks some sort of transition in behavior, operation, or state. Schumm (1979, p. 487) defined an extrinsic threshold as one that a geomorphic system will not cross unless it is driven across by a change in an external variable. A very common illustration of an extrinsic threshold would be a change of geomorphic behavior in response to climatic change. An intrinsic threshold is one that a geomorphic system may cross without being subjected to a change in an external variable. An example of an intrinsic threshold would be a change in the behavior of bedrock as it is weakened by ongoing weathering.

The schematic patterns shown in Figure 6.5 for equilibrium behavior illustrate two very important themes. First, change in landform does not preclude the possibility that the landform is in equilibrium. In fact, only in the case of static equilibrium would no change be expected, and static equilibrium is an unlikely occurrence in the natural world. Second, it is apparent that equilibrium as a concept is highly dependent upon the timescale of interest to the researcher. It is quite possible for a landform to be in equilibirum with respect to one timescale while simultaneously being in disequilibrium with respect to another.

During an equilibrium phase, or something close to it, a landform would presumably exhibit a characteristic form as suggested by Brunsden & Thornes (1979). In fact, their choice of the term "characteristic form" rather than "equilibrium form" was undoubtedly made because of the difficulty in establishing the presence of equilibrium in geomorphology. Despite the great value and significance of equilibria concepts in geomorphology, it is important to appreciate that "a tendency toward equilibrium" is much more widespread than equilibrium itself. This is because landscapes exhibit a variety of reaction and relaxation times while receiving a constant stream of inputs, a combination that all but precludes attainment of equilibrium. Indeed, this situation has been recognized mathematically by Craig (1982b) in an analysis of Appalachian landforms. Craig found that the pattern of landform behavior resembles an orbit around a central point; in other words the landscape did not approach some stable point (form), but rather circled a central point (form) with periods of near-stability inevitably being followed by periods of greater divergence. This implies that there is no unique final landform, because as one landscape unit approaches stability, it axiomatically induces instability in a neighboring unit. Similar ideas have been discussed by Thornes (1983) and are presented briefly in Chapter 14.

One equilibrium concept that is presently favored by many geomorphologists to explain landform and landscape behavior is dynamic metastable equilibrium (Schumm 1979). One facet of dynamic metastable equilibrium is complex response, which Schumm has investigated primarily in the context of river channel behavior.

If a stream is disturbed from an equilibrium state it will presumably attempt to regain equilibrium. However, this does not mean that all parts of the stream will exhibit the same trends at the same times, nor that any part of the stream will exhibit the same tendency at different times. This spatial and temporal variability in behavior as a new equilibrium state is sought has been labeled by Schumm as a complex response; it is well-exemplified in a paper by Womack & Schumm (1977).

Douglas Creek, a small stream in western Colorado, was subject to overgrazing during the "cowboy era". Since about 1882 the stream has been incising its channel bed. However, it has not done this in a consistent fashion; rather, cross sections through the floodplain (Fig. 6.6) show that incision has not only been discontinuous, but has actually been interrupted by deposition. Terraces have been dated at several locations using evidence from trees. The terraces are not only unpaired (unlike the paired terraces that would be expected from a classical case of incision), but they are also discontinuous in a downstream direction. As Figure 6.6b illustrates, the incision–deposition sequence differs markedly between cross sections; for

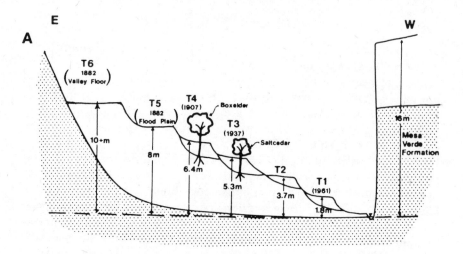

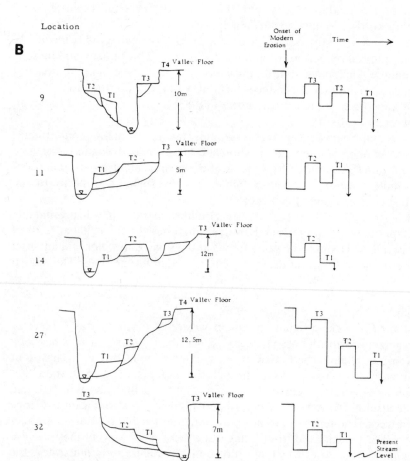

Figure 6.6.

example, at cross sections 9 and 27 there have been three depositional and four erosional phases, while at locations 11 and 32 there have been two depositional and three erosional periods. It should be noted that the terraces are numbered sequentially at each location and the numbers cannot be correlated from one location to another.

Geomorphologists focus their attention upon the morphology of the landscape. This factor, plus the difficulty of measuring energy in a geomorphic context, means that equilibrium in geomorphology is virtually always interpreted with reference to morphology and not energy. However, static equilibrium (which is perhaps the notion intuitively associated with the word "equilibrium") is only one type of equilibrium and not the one most likely to occur in nature. To make matters more difficult, equilibrium not only exhibits many forms, but it can only be identified with respect to stated time and spatial scales.

Conclusions

It is clear that geomorphic change over time is a profoundly complex issue. In examining change, geomorphologists are greatly constrained by the temporal scales at which it is possible for them to generate data. Most research projects in process geomorphology are simply too brief to establish or deny the presence of equilibrium. The multifaceted nature of equilibrium itself does much to insure this failure. Indeed, use of the term "equilibrium" without identifying which variety of it is under consideration is itself a sufficiently imprecise statement to insure failure.

Traditionally, geomorphologists have tended to interpret each landscape or landform change as a single response to a single input, i.e., a simple cause-and-effect relationship. Schumm's work has effectively put an end to this view. Unfortunately, while complex response appears to be a valid concept, it also makes interpretation of past events much more difficult. The implications of complex response for interpretation of Quaternary history, for example, are enormous because geomorphologists can no longer assign a single input to each and every response identified in the

Figure 6.6 Schumm's concept of episodic erosion illustrated by the variable number of terraces along Douglas Creek, Colorado. (a) Cross section of Douglas Creek valley at location 5. Terraces 5 and 6 represent pre-1882 valley floor and probable floodplain of Douglas Creek. Terraces 1 through 4 were formed after 1900. (b) Diagrammatic cross sections of Douglas Creek terraces. Sections are not drawn to scale. To the right of each is a representation of depositional and erosional history at each location as deduced from cross sections; that is, stream level is related to passage of time in a very general way. Note that terraces are numbered from lowest to highest at each location; therefore, 1882 valley floor is assigned a different number depending on number of terraces at each locality. (From Womack & Schumm (1977).)

stratigraphic record. In fact, there are also other (spatial) problems to overcome. However, the only conclusion that can really be drawn at this point is that research questions must be defined much more precisely temporally and geomorphologists must be much less willing to extrapolate their results if a marked improvement is to be made in understanding landform and landscape behavior over time. However, increased precision must be applied not only to time but also to the spatial scales with which it interacts; these spatial issues will now be considered.

7 Space in geo-morphology

Introduction

The Earth's rotation and orbit give diurnal and seasonal rhythms upon which our traditional concepts of time are founded. In part, this is due to their distinctiveness, but it is also due to their occurrence at a human scale – the majority of people see numerous days and many years. However, there are no similarly obvious starting points upon which we base our view of space. Clearly, we lead our daily lives within a restricted space with which we are very familiar and have larger and larger spaces with which we are less and less familiar; but none of these has a natural measure.

In the absence of natural spatial measures, we have created some; for the most part they are linear (e.g., the meter and kilometer, the yard and the mile), although there are areal measures (e.g., the hectare and the acre). It takes only a moment's reflection to appreciate that the absence of natural spatial measures has resulted in a much greater variety of man-made ones being created than is the case for time.

Geomorphologists are faced, therefore, with no obvious starting points upon which to found their concept of space. Nor, indeed, are there any obvious appropriate or correct spatial scales at which geomorphological research should be conducted. While it might be argued that a slope profile represents the fundamental linear measure, and a drainage basin the fundamental areal measure, both exhibit great variation in size and offer no realistic foundation for a formally defined yardstick.

Haggett et al. (1965) made an attempt to establish a formally defined areal measure in geomorphology by suggesting that the Earth itself makes an obvious starting point and that spatial scales may be founded upon systematic subdivision of this fundamental unit. The suggestion has logic on its side, but lacks anything to which we may easily relate and failed to achieve any widespread support.

While there are no readily apparent natural measures with which to organize spatial issues in geomorphology, there are other fundamental concepts that we apply to spatial issues. Nystuen (1963) attempted to identify the primitive (undefinable) concepts or terms that underpin the discipline of geography; his findings are equally applicable to spatial issues in geomorphology. Nystuen suggested that directional orientation, distance, and connectiveness form the three concepts from which all of our other spatial ideas are derived.

Orientation is an obvious enough property which may be applied to linear features. When an object itself has no orientation, Nystuen (1963, pp. 378–9) noted that the direction between two objects becomes the issue. Distance is usually viewed as the shortest distance between two objects by whatever measure is being applied. It must not be forgotten that distance may be measured by such things as travel time as well as by the more familiar kilometers or miles. Nor can it be forgotten that the distance between two objects (e.g., A to B, B to A) may not be symmetrical (downstream flow in a river channel being a case in point). Connectiveness excludes concepts embraced by directional orientation and distance, and may be most readily envisaged as relative position (contiguity and adjacency are other synonyms suggested by Nystuen (1963, p. 379)).

In the same way that we are unable to examine temporal issues meaningfully without defining a spatial scale, so we find time intruding into any consideration of spatial topics. Nystuen (1963, p. 382) highlighted the importance of "historical tension" in spatial issues; by this expression he simply meant that past events derived from processes no longer functioning may play an important role in controlling contemporary spatial patterns. This is certainly very true in geomorphology where the slow rate of geologic and geomorphic change may create profoundly important constraints on contemporary processes and ensuing forms.

So great is the influence of the past in geomorphology that Chorley (1978, p. 8) noted that: "During the 1960s it became clear that the majority of mesoscale landform assemblages represent a palimpsest of superimposed and interlocking process–response systems of highly varying relaxation times." This statement introduces two terms that have become popular in geomorphology. Micro-, meso-, and macroscale have become widely used to mean very small, medium-sized, and large scales, respectively, although none has any formal definition. Palimpsest has also become a popular descriptive term. Originally it was used to describe a piece of parchment that had been written upon more than once. Parchment was so valuable that once used it would be erased and written upon again. However, the erasures were rarely complete and earlier inscriptions, as well as the current ones, could be deciphered. By extension, geomorphologists have taken up the term simply to mean that the contemporary landscape bears the imprint of earlier processes (regimes) as well as present-day ones. The extent to which a geomorphologist is likely to identify residual effects will be dependent very much upon the model of landscape evolution he or she uses. The nature of these various models is the subject of Chapters 10–13, while the survival of ancient landforms (paleoforms) has been discussed by Twidale (1976).

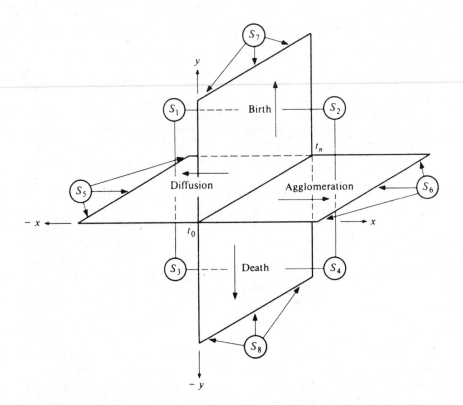

Figure 7.1 A diagrammatic view of spatial processes. The diagram was created to view human processes – accordingly, the geomorphologist must make some adjustment in terminology. Three axes are depicted: the horizontal x axis represents the diffusion–agglomeration continuum; the vertical y axis represents the birth–death (initiation–elimination) continuum; the t axis represents time from t_0 (earliest time period) to t_n (final time period). The following summary (after Getis and Boots 1978, pp. 4–7) of the spatial process in each sector assumes that x and y are constant over time.

Sector S_1 $(-x, y)$: births exceed deaths, and diffusion exceeds agglomeration. S_{1a} exhibits the "contagion effect" – the number of instances increases through time and at locations further away from the original source. S_{1b} exhibits the "Pied Piper effect" – the number of instances increases through time and are relocated.

Sector S_2 (x, y): births exceed deaths, and agglomeration exceeds diffusion – a "family development effect".

Sector S_3 $(-x, -y)$: deaths exceed births, and diffusion exceeds agglomeration. S_{3a} is the "survival-of-the-fittest effect" – the number of objects decreases through time and diffuses. S_{3b} is the "lost-cause effect" (opposite of the "Pied Piper effect") – a group moves, but loses numbers.

Sector S_4 $(x, -y)$: deaths exceed births, and agglomeration exceeds diffusion – numbers decrease through time while clustering continues.

Sectors S_5 $(-x, 0)$ and S_6 $(x, 0)$: no increase or decrease in numbers over time. In S_5 migration dominates, while in S_6 there is a "town-meeting effect".

Sectors S_7 $(0, y)$ and S_8 $(0, -y)$: pure birth and death respectively with location being derived from a purely random process. (From Getis & Boots (1978).)

The complexity of space–time interaction, which must be unraveled if truly spatial issues are to be studied, was shown schematically by Getis & Boots (1978, p. 4), and is shown here as Figure 7.1. Like Nystuen, Getis & Boots were primarily interested in human issues where human mobility has led to much greater theoretical refinement concerning spatial matters than is generally demonstrated in geomorphology. The three axes defined by Getis & Boots have obvious applicability to geomorphology. Similarly, some of the effects they suggest have obvious geomorphic counterparts, although others would appear to be of limited applicability in geomorphology.

Even when temporal contamination has been removed, examination of spatial issues still includes many difficulties. Nystuen (1963, p. 382) pointed to the importance of "dimensional tension" – interaction between features that are points, lines, and areas. This is certainly something of fundamental importance in geomorphology and may be easily highlighted by considering the relationship between a small feature on a hillslope (effectively a point) and its relationships to the local fluvial network (a linear feature) and to the drainage basin (an area) within which it falls.

One of the most elemental issues to pervade geomorphology is the spatial attribute of size or scale. Size alone plays a profound role in the behavior of both individual landforms and landscape assemblages. It is very apparent that geomorphologists have yet to come to grips fully with this issue, although some of its salient aspects are at least recognized. Among the most important themes to emerge from theoretical considerations of scale are scale coverage, scale linkage, and scale standardization (Haggett 1965).

Spatial issues will be reviewed by considering Nystuen's three primitive concepts individually, followed by a brief comment on the importance of boundaries, and concluded with an examination of Haggett's three scale themes. Shape or morphology, which is the spatial issue that has preoccupied geomorphologists longer than any other, will be treated separately in the next chapter. The techniques that have been employed by geomorphologists to study spatial topics will not be reviewed. There are several good reviews of individual analytical techniques, including Cole & King (1968), Doornkamp & King (1971), Chorley (1972), and Craig & Labovitz (1981). This list of texts is by no means exhaustive, but does illustrate the diversity, rapid evolution, and growing sophistication of the tools used by geomorphologists to analyze spatial data.

Orientation

A fundamental attribute of any geomorphic feature is the direction in which it faces (commonly called its aspect). This may be seen in any number of contexts. For example, in a terrestrial situation the geomorphic regimes of north- and south-facing slopes may differ quite profoundly in a single region. Similarly, in a marine context shoreline orientation may be of fundamental importance through its influence on wave fetch, as well as tidal and longshore processes. While secondary in most instances, it is important to remember that some geomorphic components may also have orientation in the vertical, rather than the horizontal, plane, an example being imbricated, river-bed cobbles.

The most important issue in analyzing orientation data is their occurrence on a circular scale, which makes definition of a zero value both arbitrary and profoundly important. This may be illustrated by considering the mean orientation of two directional values, 1° and 359°, which is 180°. Such a mean value runs counter to intuition and is clearly completely inappropriate from our normal viewpoint of directional data. Fortunately, circular distributions have been the subject of considerable attention and, while much less well-known than their straight-line counterparts, there is a fully developed set of statistical techniques for analyzing them; Mardia (1972) provides a comprehensive starting point.

In geomorphology there is relatively little use of orientations derived from joining point features by a line (one of Nystuen's orientation concepts). But it is a technique that finds occasional application. Some examples are the following: the northwest–southeast orientation of the Hawaiian Islands in which island age increases to the northwest; the relationship between a sediment source, prevailing winds, and a dune field (e.g., White Sands, New Mexico); the distribution of glacial erratics with reference to the parent bedrock outcrop to determine direction of glacial flow. However, there is no widely used research technique in this sphere beyond such straightforward source–resultant feature relationships.

A much more frequent use of orientation in geomorphology is analysis of the orientation of linear and/or areal features. The distinction between linear and areal often depends upon the scale of study; for example, a ridge may be considered linear at a small scale, but areal at a large scale. Geomorphic orientation is often directly inherited from geologic inputs (fault, fold, outcrop strike, regional dip), but may also be indirectly associated with geologic inputs (e.g., a fault-line scarp).

Complex interactions involving directional attributes are fairly common in geomorphology, especially in features that respond quickly (i.e., have short reaction and relaxation times). Orientation of thaw lakes in places such as the Arctic Coastal Plain of Alaska (see Washburn (1980, pp. 271–3) for a brief discussion) represents a still uncertain set of interactions. Better-understood relationships exist between prevailing winds, sediment sources, and sand dunes (e.g., Harris 1974). Many such relationships exist in coastal contexts where orientation of the coastline, the variety of wave trains, and tidal influences combine with sediment source locations to produce an enormously complex geomorphic environment with, in the case of beaches, for example, remarkably short reaction and relaxation times.

Dip, or orientation in the vertical plane, is a physical attribute widely used in geology. It has considerable utility in geomorphology as well, but is generally less widely studied. Most commonly it has been used to interpret flow processes: in fluvial contexts, where cobble imbrication may be a useful measure (e.g., Richards 1982, pp. 120, 192–3); in glacial environments, where till fabric analysis has been a widely employed diagnostic technique (e.g., Andrews 1971, Mark 1974); as well as in sand dunes (e.g., McKee 1979).

Distance

The separation of objects or features, which is what we are generally measuring with distance, is quite obviously a fundamental property of their relationship. In most instances in geomorphology interest centers on relationships between things that decay or are reduced as the distance between them increases. However, other distance relationships are feasible, if less common. It is possible for something to be uniform over a given distance until a boundary is reached; it is also possible for something to exert an influence only at a distance. In nearly all instances distance between a cause and an effect (or response) is being measured; therefore, the distance is asymmetrical because it is of significance only in one direction.

Spatial decay functions abound in geomorphology. In the eolian realm, point sources (a volcano), linear sources (a beach), and areal sources (sandar) may all be related to eolian deposits (ash, dunes, and loess, respectively) by decay functions. In general, the distance decay incorporates changes in both quantity and particle size; Ruhe's (1969, pp. 29–37) classic work on Midwestern loesses serves as an example. However, an important complication must be borne in mind and that is that secondary reworking of old deposits is widespread. Thus, volcanic ash and loess may be deposited as blankets across an entire landscape; however, they often only survive as discontinuous, reworked deposits. In such cases they are concentrated in locally favored depositional microenvironments and their thickness in these locations does not reflect initial depth of deposition.

In human affairs it is relatively simple to identify areas over which some attribute is unvarying; Nystuen (1963, p. 379) pointed to the rule of law over political units. However, such situations are not nearly so obvious in geomorphology. It is clear that in many instances where a geomorphologist identifies regions or zones he/she treats them as if some attribute within them is unvarying; for example, when a specific type of landform assemblage is associated with a particular rock type. This is often done although it is known not to be strictly true. Even where detailed studies recognize local variability, such as facies or rock thickness changes, by further subdivision, this does not change the principle but only the scale of resolution.

It may also be possible, in very restricted circumstances, for cause and effect to be separated by distance over which the effect is absent. The most important examples in geomorphology would seem to be coastal, and stem from the ability of water to transfer energy in wave form with little loss in deep water. Upon entering shallow water the energy may be released violently and locally having left no mark over the intervening distance. The premier example in this category is the tsunami.

All that has been written about distance with respect to the horizontal plane may be repeated with reference to the vertical plane. However, this is generally of lesser importance in geomorphology and is invariably restricted to much shorter distances. Nevertheless, the importance of vertical distance decay from source to result cannot be ignored. Unloading in the case of bedrock weathering and depth of permafrost serve as simple, but important, examples.

Connectiveness

Relative position is a concept that may be applied to point, linear, and areal attributes. In fact, all three types have received theoretical attention in geomorphology, although not with the same intensity as in human geography.

In their basic diagram Getis & Boots (1978) indicated that agglomeration and diffusion (dispersion) are the two basic opposing spatial tendencies of point patterns (see Fig. 7.1). This viewpoint, which unifies all patterns within a single concept, was also discussed by Dacey (1973) and is illustrated in Figure 7.2a. In this approach a random pattern is considered to be a midpoint between the opposing forces that create clustered and regular patterns.

Point patterns may also be viewed as exhibiting independent characteristics (Fig. 7.2b); Dacey (1973, p. 135) also discussed this approach, defining the independent concepts as:

(1) The *pattern* of a spatial distribution is the areal or geometric arrangement of the geographic facts within a study area *without regard* to the size of the study area.
(2) The *density* of a spatial distribution is the overall frequency of occurrence of a phenomenon within a study area *relative* to the size of the study area.
(3) The *dispersion* of a spatial distibution is the extent of the spread of the geographic facts within a study area *relative* to the size of the study area.

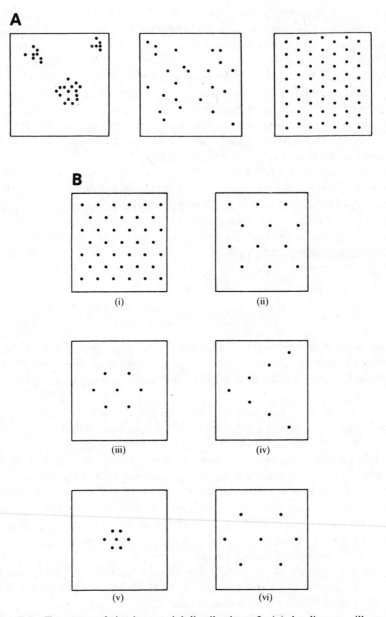

Figure 7.2 Two ways of viewing spatial distributions. In (a) the diagrams illustrate (from left to right) clustered, random, and regular spatial distributions. In (b), diagrams (i) and (ii) have the same patterns and dispersions but differ in density; diagrams (iii) and (iv) have the same densities and dispersions but differ in pattern; and diagrams (v) and (vi) have the same densities and patterns but differ in dispersion. (From Dacey (1973), part (b) being taken by Dacey from Thomas (1965).)

Without belaboring the relative theoretical strengths of these two approaches, they clearly establish the presence of an adequate theoretical base against which such patterns may be measured. However, as Getis & Boots (1978, pp. 14–15) emphasized, pattern analysis is a useless undertaking unless there is a process-based hypothesis postulated prior to pattern measurement.

Examples of geomorphic research in which pattern analysis has formed the foundation are not that plentiful. Presumably this is because most geomorphic features are considered as areas, or where they may be effectively treated as points there has been a tendency to focus upon individual genesis rather than population pattern. McConnell & Horn's (1972) study of sinkhole distribution on the Mitchell Plain of Indiana serves as a good illustration of the manner in which such a study should be approached.

Linear patterns, or networks, have been more widely investigated in geomorphology because stream networks, along with other fluvial topics, have long been central to the discipline. Stream network analysis was one of several topics introduced to geomorphology by R. E. Horton (1945) and further developed by A. N. Strahler and his students at Columbia University.

One important revelation from stream network analysis was identification of the importance of randomness in stream networks (see Strahler (1969, pp. 481–500) for a simple but comprehensive summary). In some ways this thwarted further development of the topic, but it is a research topic that is still pursued by a small group of fluvial geomorphologists. From the perspective of the present text the work of Shreve (1966, 1967, 1969) is particularly interesting because he specifically attempted to compare natural stream networks against theoretically determined patterns.

Areal concepts are often included within morphometric studies, but will be considered briefly here. Two separate issues may be distinguished: (1) consideration of the two-dimensional shape of individual landforms; (2) examination of the manner in which groups of forms cover areas larger than a single individual form. Obviously, the former does not fall logically into the present topic; however, it deserves brief mention because of implications for the matter at hand. The bulk of areal research in geomorphology has focused upon drainage basins (often considered the fundamental areal geomorphic unit). Individually, their shape has been compared to that of a circle (e.g., Gregory & Walling 1973, pp. 51–7). Such an approach makes good sense because, in the two-dimensional case, a circle represents the smallest perimeter-to-area ratio and consequently is the lowest free-energy form. Nature, in general, is commonly assumed to have strong tendencies toward lowest free-energy conditions.

When considered as assemblages, circular forms are not the most efficient ones with which to cover entire surfaces comprehensively. This is for the obvious reason that circles cannot be packed so closely together that their boundaries are in continuous contact with each other. The geometric form that provides the best total coverage of a surface is the hexagon (a six-sided figure). Thus, there is a sound theoretical foundation from which to evaluate areal coverage by some sort of repetitive feature that would be expected to cover an entire surface. At the drainage-basis scale, Woldenberg (1972) provides a good example of this type of research. However, it may be seen at many other scales in nature ranging from the cracks in dried mud, through columnar jointing in basalt, up to patterned ground phenomena (see Washburn (1980) for a discussion). In general, such studies have been fairly weak with respect to relating areal form to genesis, but Ray *et al.* (1983) provided an interesting example in which patterned ground regularity, including depth–width ratios, is related directly to the formative processes.

Boundaries

Boundaries are not specified by Nystuen as a fundamental spatial property, but it is important to appreciate that in practical terms they are an important spatial issue. This is illustrated by Dacey's (1973) definitions of density and dispersion, both of which incorporate boundaries, and again in Woldenberg's (1972) paper in which he attempted to demonstrate that surfaces may be subdivided into nested sets of hexagons.

Identification, demarcation, and justification of boundaries is extremely problematic in geomorphology. There are at least three widespread contexts in which boundaries have been used in geomorphology. First, there are those boundaries associated with traditional regional studies (e.g., Thornbury 1965). Such boundaries are generally very fuzzy and have rarely been defined in anything approaching a formal sense. Not uncommonly these boundaries include mixed criteria and simply do not stand up to theoretical scrutiny.

A second type of boundary is that associated with identification and delimitation of drainage basins (i.e., the watershed). This kind of boundary may be approximated quite readily and is important because drainage basins are commonly treated as if mass cannot cross their boundaries. The primary difficulties with this approach have stemmed not from boundary definition but from assumptions that there is complete geomorphic linkage within the basins (Costa (1975) provided a good illustration that this is often not true) and that materials being exported from the basin via its stream(s) represent denudation (surface lowering). In fact, sediment and solute export from the basin may well reflect for the most part simple reworking of floodplain alluvium (e.g., Costa 1975, Trimble 1975, 1977).

The third boundary context includes all situations where a discrete landform is identified and modeled. The modeling process is often pursued in process geomorphology using systems techniques (see Ch. 12) and boundary definition should be noted as a primary limitation on systems modeling. Even when systems techniques are not used, identification of boundaries delimiting discrete landforms is an extraordinary, rather than an ordinary, undertaking whose difficulties are well illustrated in Evans & Cox (1974).

Scale coverage

In practical terms the central issue in scale coverage is that of sampling. It is a topic about which reams have been written and an exhaustive review is beyond the scope of this text; comprehensive starting points may be found in Gregory (1978), Krumbein & Graybill (1965), Till (1974) or Yeates (1974). In geomorphology much sampling is purposive, rather than random. This is often the case when a particular form of relatively limited occurrence is the focus of interest. As a result many statistical summaries in geomorphology papers are best thought of as descriptive statistics rather than inferential statistics, i.e., they describe the attributes of the group measured rather than providing a mathematically sound estimation of the entire population.

One of the most critical problems encountered in any sampling design is that of spatial autocorrelation (Ebdon (1978) provides a good basic explanation). As each measure in a sample is supposed to be independent, serious violations of sampling theory occur if, in reality, neighboring values influence one another; this is often the case in geomorphology and is the basis of spatial autocorrelation. Relatively few papers in geomorphology formally consider the issue of appropriate sampling intervals; Caine (1982) provided an interesting *post hoc* review of the topic with particular reference to research plots designed to monitor surficial movement on alpine surfaces. Caine's results suggest that research plots should be at least 12 m apart to insure independence; however, this is undoubtedly a scale-dependent value and, therefore, the 12 m spacing would only be appropriate for research plots of about 1 m^2.

Spatial autocorrelation is often an important problem, but it is also an attribute that may be exploited. By varying sample intervals and monitoring the behavior of spatial autocorrelation coefficients, it is possible to investigate the wavelengths (i.e., the pattern of topographic highs and lows) that are characteristic of particular landscapes; Craig (1982a, b) provided examples of this type of research. Mark (1975), in a discussion of morphometry parameters, formally defined selected wavelength parameters as a basis for numerical landscape description. The primary difficulty in this type of approach in process geomorphology, as opposed to static or morphometric geomorphology, is the extraordinarily high data generation requirements. Nevertheless, it is an issue that must be addressed sooner or later.

Another important aspect of sampling in process geomorphology is stratification of the landscape. If step-function patterns such as those envisioned by Schumm (1979) are a reality, and if the landscape is to be modeled as systems and subsystems as suggested by Chorley & Kennedy (1971), it is not appropriate to sample at random without first identifying the different subunits present. Stratification in geomorphology is probably best exemplified by the use in hydrology of unit-source areas.

Hydrologists recognize that water yield to a river varies dramatically within a drainage basin. Many factors influence this situation: cover type, soil depth, position on the slope, and position within the basin, to name but a few. Accordingly hydrologists stratify a basin into homogeneous units called unit-source areas (Amerman 1965), such units being assumed to have the same hydrologic response. A subset of each type of area may then be monitored (sampled) and these instrumented areas are called unit-source watersheds. By making really weighted calculations, a much more precise model of basin-wide hydrologic behavior may be generated. Such dynamic watershed models are now extremely sophisticated and widely used in flood forecasting and water management. This still does not address the problem of partial areas within a unit-source area or basin (Weyman 1974). Here the problem is that the area actually contributing runoff varies with time during a storm, and also varies from storm to storm depending upon antecedent conditions.

In a similar fashion geomorphologists may attempt to identify unit-source areas with respect not only to hydrology but also to closely akin fluxes such as solute load, as well as to highly disparate ones such as fluxes of coarse detritus or eolian sediments. Of course this is extremely complex and has yet to progress very far. There have been one or two attempts to render selected mass wasting fluxes into comparable units, notably Rapp (1960) and Caine (1976), but these studies did not examine unit-source areas.

Generalized surficial movement as monitored by sediment traps has been categorized by Bovis & Thorn (1981) on an alpine interfluve according to vegetation communities. This approach is essentially a unit-source area one, but requires a large-scale vegetation map. Where sediment trap data and vegetation mapping may be combined successfully, geomorphic results can be highly revealing. Bovis & Thorn (1981) found order-of-magnitude differences in surficial soil yield between different alpine tundra plant communities.

Subsequently, Thorn (1982) expanded the research to examine the impact of pocket gopher burrowing over the same area and, once again, order-of-magnitude differences and a distinct step-function emerged between alpine plant communities. However, there are important limitations to such work. In the case of the papers by Bovis & Thorn and Thorn it is unlikely that the surficial processes monitored are significant in long-term landscape evolution; nor did the surficial sediment yield and gopher burrowing maxima coincide. Finally, it should be appreciated that variation in vegetation type is recognized on alpine tundra merely as a convenient surrogate measure for seasonal snowpack distribution.

Scale linkage

The scale-linkage problem stems directly from practical constraints that force a researcher to measure or monitor on a manageable (small) scale even when interested in much larger populations or areas. Conceptually, it is simply recognition that scale alone is a significant variable and that the results of research undertaken at one scale are not necessarily applicable to another scale. Three strategies to overcome this problem in geomorphology are worth mentioning.

One possibility is to use a methodology that specifically excludes scale variability. The use of ratio data is an integral part of many Hortonian parameters and papers by Schumm (1956) and Strahler (1957, 1958) are good examples of the attempt to overcome scale variability by using ratio or dimensionless values. Early success with this approach has not produced sustained interest. Once randomness was recognized as the salient characteristic of Hortonian distributions it, in turn, had to be explained. "Random" is, in fact, a rather elusive concept (Getis 1977) and not easily defined. However, we may distinguish between random processes and random patterns (distributions) and should appreciate that a random pattern may not result only from random processes. Consequently, a random pattern is but one example of equifinality (i.e., it may arise from different individual processes or mixes of processes).

Another widely used approach is to smooth locally generated data so that broad patterns are emergent. In geomorphology this methodology has been dominated by the use of trend surface analysis, supported by an array of secondary techniques that permit detailed extraction of the characteristics of contributing components and of significant deviations therefrom. Again the texts on this topic are legion and cannot be adequately summarized here; however, Davis (1973) provides some good introductory reading.

A third possibility that has not been used so widely is to synthesize landforms by reformulating them using data derived at a smaller scale. Caine (1979) collected surface process data from 1 m^2 study plots, and then clustered the plots using factor analysis. His next step was to place the averaged plot data in correct sequence with respect to both distance from the stream and aspect. The net result is a synthetic cross section that may be viewed as a cross-valley or cross-interfluve section depending on where the sequence is broken. Burns & Tonkin (1982) used a similar, though less refined, approach to produce a synthetic alpine slope model for alpine soils. By examining soil attributes in a variety of alpine interfluve contexts they were able to produce what is essentially a set of catenary relationships; however, the postulated catena is synthetic in that the entire sequence has never been examined or observed down a single slope profile.

This synthetic approach is appealing because it enables the researcher to recombine data gathered at one scale into meaningful representations of larger-scale units while retaining statistical information. It also has the virtue, unlike standard statistical abstraction, of producing something that may be readily envisaged and checked physically. Finally, these strengths may be combined and the approach seen as an excellent one in hypothesis formulation.

Scale standardization

Given the reality that spatial scale is itself an important source of variability it is reasonable to expect that geomorphologists would have devoted considerable time to scale standardization, but this has not been the case. One very generalized standardization has been the focus upon specific landform units, e.g., drumlins or inselbergs, so that some natural constraints have been embraced; however, many natural phenomena occur over enormous scale ranges. The tendency towards meso- and microscale research in process geomorphology has also been quite clearly driven by logistical constraints. However, this constraint or limitation has not led to standardization, the bewildering array of plot sizes used worldwide in soil erosion studies being a classic example of this failure.

In process geomorphology most rate measurements are effectively point data. This should not be allowed to mask the incompatibility in results stemming from variations in plot or trap size, manner of installation, and monitoring duration and frequency. Nevertheless such sources of variability are probably secondary to the errors stemming from extrapolation of one or two years of data over geomorphically useful timespans, which range from hundreds to, more commonly, thousands of years.

Investigations of process mechanisms are most frequently attempted in the laboratory. In these circumstances the scale of investigation is constrained primarily by technical considerations generated by instrumentation capabilities. Therefore, detailed analysis often makes very small-scale experiments axiomatic. This brings to mind Chorley's (1978, p. 10) analogy of the different skills required of the traffic engineer and the car mechanic. Is it really relevant or feasible for geomorphologists to examine pore pressures across a membrane in a laboratory and expect to extrapolate these findings to hillslope behavior? The alternative approach in the laboratory is to scale down large segments of the landscape. Again interesting insights may be obtained, but in this instance incompatibilities between scale modifications are created and must be recognized (King 1966, pp. 188–229).

Recognition of scale standardization as a focal problem is merely the first, halting step of rectification. Investigations and derivation of appropriate scales should be primary objectives in geomorphology but clearly are not, except for a very few researchers. Hierarchical nesting of plot, trap or other data-generating techniques, together with examination of the patterns of behavior of associated correlation and/or variance (basically the approach of a botanist to species diversity, or that of Craig cited earlier in the chapter), is clearly critical. However, while this approach is labor-intensive, but feasible, with static measures it will be inordinately time-consuming for process-rate studies.

Considering research-plot data as point data, virtually no standardization has been achieved even at national levels, let alone international levels. If attention is transferred to linear topics, the most widely studied geomorphic theme is the slope profile. Here there is a very broad standardization; for example, virtually all studies extend from ridge crest to the basal stream. There have also been some other more stringent attempts to standardize slope profiles; for example, recognition of main slopes (Carson & Kirkby 1972, pp. 377–89) and restriction of studies to them alone and similar restriction to straight slope units (Carson 1967). However, even these limitations are rather general and also exclude many interesting landscape components. Areally, use of Horton's drainage-basin ordering scheme (most frequently Strahler's (1952b) modification thereof) has produced a very broad degree of standardization, as comparisons are generally only made between basins of the same order. However, what physical significance this has is certainly unclear. Furthermore, not only do basins of the same order vary enormously in size, but they may vary greatly in age, and, when basins of differing order are considered, time may well be the salient variable.

Conclusions

It is clear that the biggest difficulty in evaluating the role of space in geomorphology most frequently stems from its inextricable entwinement with temporal influences. There are exceptions to this statement but they invariably involve embracing characteristic-form (Ch. 5) or time-independent (Ch. 11) viewpoints that have limited use. This is not to suggest that such approaches are without merit because both Kirkby's (Ch. 13) mathematical modeling and Hack's more general statement of time independence (Ch. 11) are extremely productive under some circumstances. However, such ideas appear to be most successful with smaller-scale landforms, and may well be intrinsically inapplicable to entire landscapes (e.g., Craig 1982b).

An alternative method of extracting temporal contamination from an investigation of spatial issues would be to stratify subject matter temporally. It is unlikely that this could ever be done with a high degree of precision, but the availability of numerous isotope measures ranging over most of the timespans of interest makes such an approach feasible in some circumstances. Vita-Finzi (1973) forwarded a persuasive argument for putting much greater emphasis on isotope dating, particularly ^{14}C.

The temporal admixture (Nystuen's historical tension or the palimpsest concept) is closely followed as a source of difficulty by a purely spatial issue, that of scale linkage in which size alone is recognized as an extremely influential variable. If, for the sake of discussion, it is assumed that all temporal influences could be removed successfully, then something like static allometry might be used to investigate purely spatial influences on scale linkage. It is also possible that dynamic allometry might be useful in some instances, although slow rates of change are likely to render this approach infeasible for large landforms.

Dimensional tension as envisaged by Nystuen is an all but unknown topic in geomorphology. However, it is one that lies at the heart of such issues as drainage-basin studies. If drainage basins are not well-integrated geomorphologically (and this seems likely, despite widespread assumptions to the contrary for many years), it will be essential to investigate the relationships between river channels and interfluves, via the floodplain, before any real progress can be made. This certainly is a dimensional issue at this scale as the linear channel is integrated with the areal basin, or sections thereof.

Boundaries remain as another profound problem in geomorphology, but one that is certainly no greater than in other disciplines. They are, of course, a purely human problem stemming from an intellectual need for simplification that mandates division of the indivisible. As such it is unlikely that there will ever be a truly satisfactory theoretical basis for performing the task, merely more widespread appreciation of the difficulty of undertaking it and of the compromises always associated with it.

Nystuen's three primitive spatial concepts, direction, distance, and connectiveness, seem more than adequate for our current understanding of spatial issues. Similarly, Haggett's identification of scale coverage provides an adequate reminder, but still a frequently overlooked one, of the needs of sampling. But his recognition of scale standardization as a critical issue is something that geomorphologists have either forgotten or chosen to ignore. It is not a simple issue, and the availability of right and wrong scales is obviously a misconception. Nevertheless, geomorphologists must pay more attention to standardization because it is certain that many assumptions concerning both agreement or apparent disparity between individual studies are man-made artifacts stemming directly from scale divergence.

Many of these spatial issues have effectively been placed on hold, or the backburner, by a discipline-wide preoccupation with genesis or change over time. This has resulted in a very strong emphasis, in process geomorphology in particular, on developmental sequences of individual landforms. Such an approach is not intrinsically bad or wrong, but it does seem that geomorphologists now have sufficient evolutionary information of many individual landforms to form an adequate starting point for reconsideration of the behavior of assemblages and their interactions.

8 Morphology

Introduction

The majority of sciences have begun with careful observation and rapidly included descriptions of shape or form as a basic ingredient. Geomorphology is no exception and the centrality of land*forms* to the discipline at all stages of its development, including the present, bears witness to this. As a simple verification of this claim, consider the role played by field sketches in early geomorphic publications, the role of photographs in today's publications, and the pre-eminent role of the slide show in contemporary oral presentations.

Given the great significance attached to form, it is appropriate to attempt to define it and also to consider why it is so important. Surprisingly, there is no formal definition of what constitutes a landform in standard geomorphic references such as *The encyclopedia of geomorphology* (Fairbridge 1968, pp. 618–25) or the *Glossary of geology* (Bates & Jackson 1980, p. 349). In both instances the reader is given commonsensical, narrative descriptions that are essentially identical to those found in a standard dictionary, but applied to geomorphology specifically, and encompassing the entire range of scales at which geomorphologists work. In most dictionaries and thesauruses "form" and "shape" appear as synonyms and are defined something like "the quality of a thing that depends on the relative position of all points composing its outline or external surface" (*Webster's* 1977, p. 1667). In fact, external form alone does not fully describe what most geomorphologists understand by the term "landform" because, as noted by Haines-Young & Petch (1983, p. 466), structural (and other internal) attributes, as well as functional roles, are often incorporated in descriptions/definitions of landforms.

While a commonsensical, narrative-style definition of form or shape will serve many purposes, it will not suffice for many other worthwhile undertakings. In virtually all instances geomorphologists have been and are interested in landforms not simply for their own sake but, rather, to gain understanding of the processes that have produced them. Furthermore, this interest in formative process is not always the ultimate goal, but itself may only be an intermediate step in reconstructing paleo-environments. Once interest shifts to relationships between process and form, the door is opened to a seemingly unending quest for increased precision. The root of this tendency lies in the fundamental conceptual requirement that, if form is to shed light on process, there must be a unique, one-to-one correspondence between a process and the ensuing form. If this condition does not pertain then the study of form is not nearly so appealing as a research methodology.

In fact, it has usually been widely held in geomorphology that a given landform may attain its shape through a variety of nonsimilar developmental pathways. This is the concept of equifinality, or convergence in the European literature. Haines-Young & Petch (1983) discussed the issue of equifinality at some length; salient among their conclusions were that equifinality is frequently identified as the result of imprecision, rather than being truly present, and also that equifinality is more usefully defined as a situation in which the same landform emerges as the result of the action of the same process(es) upon landforms of differing initial conditions. The latter situation has been illustrated in geomorphology by Kirkby's (1971) work on hillslope profiles under specific process regimes.

If the theme of one-to-one correspondence between process and form is pursued, it follows that landforms should, ideally, have brief reaction and relaxation times because process–response (form) modeling is dependent upon characteristic-form concepts. If the logical alternative is assumed, that reaction and relaxation times are long, then it is extremely difficult to establish that there is a functional relationship between contemporary processes and forms. The study of form then becomes an end in itself with little or no theoretical underpinning.

If equilibrium between present-day landforms and processes is assumed, then the tasks of retrodiction (the traditional goal) and prediction (a new, but growing trend) become conceptually very narrow, if not impossible. This stems from the fact that, with short reaction and relaxation times, long-term retrodiction and prediction require unchanging conditions. The only logical way out of the dilemma posed by either long or short reaction and relaxation times is an ergodic assumption in which space is substituted for time.

Logically, yet another perspective must be included and that is the notion that, once an equilibrium or characteristic form is created, it can only be destroyed, but cannot be transformed into a second characteristic form. At first sight this may seem indefensible, but it describes a fairly common position taken by geomorphologists with respect to both erosional and depositional forms. Geomorphologists are prone to regard many landforms as degraded examples of some conceptualized, fully developed landform. In fact, this is a timescale issue and embraces recognition of lengthy reaction and relaxation times for many landforms. It is a view that may be sustained if there has been some profound shift in environment between the constructive and destructive phases of the landform. Examples of such situations would include sub-aerial erosion of a lacustrine delta (e.g., some aspects of the work by Ryder (1971)) or some aspects of sub-aerial weathering of granitic forms as discussed by Twidale (1987). Nevertheless, any such view is always jeopardized by the possibility that the so-called degraded feature never attained full development, or that full development is extremely variable in form.

Such issues as precise definition of form and shape have raised much more interest in some other branches of Earth science than they have in geomorphology. In sedimentology intense interest in using the outward nature of particles to interpret formative and depositional environments has led to form, shape, and related concepts receiving much attention, including attempts at formal definitions (e.g., Griffiths 1961, Whalley 1972, Barrett 1980).

In summary, it is clear that the issue of form is central and fundamental to geomorphology, but poorly defined by contemporary standards in process geomorphology. The fundamental split in form studies is between traditional, descriptive (qualitative) approaches and quantitative descriptions. Either may be wedded to some sort of theoretical foundation, but either may also lack a theoretical underpinning. Two approaches to the landform issue have been morphological mapping and morphometry. Each of these two approaches will be reviewed briefly and the development of morphometric parameters considered by examining a linear, an areal, and a three-dimensional example.

Morphological mapping

Ollier (1977, p. 277) observed:

> From such ideas the concept of terrain classification has arisen, that all landscapes can be divided into smaller units. Some units may be unique (a meteor crater, for instance) but most will be made up of a number of repeated landforms, and the landforms in turn will consist of assemblages of still smaller landscape units such as ridge tops, midslopes, valley floors, etc.

Clearly then the objective of morphological mapping is to recognize basic units that are easily identified in the field, on aerial photographs, or from maps. The task seems simple enough at first sight but when operationalized becomes surprisingly difficult.

A fundamental unit is to a large degree a scale-dependent concept. A ridge may be viewed as a single unit in a regional context but as a multiplicity of units in a slope profile. So scale and objective quickly become deciding factors in acceptance or rejection of morphological mapping.

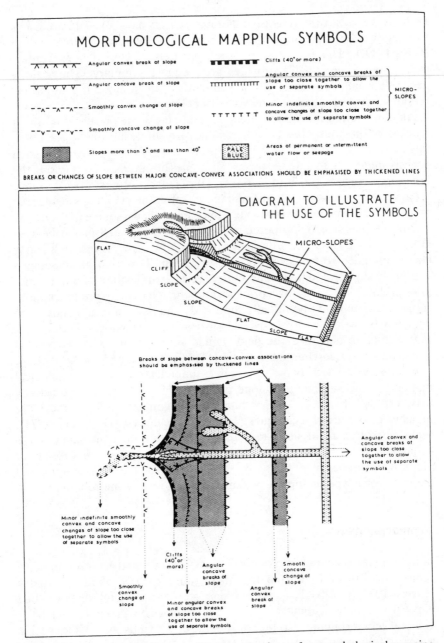

Figure 8.1 A schematic diagram of Savigear's scheme for morphological mapping. (From Savigear (1965).)

Perhaps the most comprehensive hierarchical scheme is that of Australia's Commonwealth Scientific and Industrial Research Organization (C.S.I.R.O.). The definitive outline appeared in Christian & Stewart (1953), but Ollier (1977) provides an adequate summary. The scale is generally small, or coarse, by the standards of most modern geomorphic research, and the objectives are clearly applied.

An objection to much morphological mapping is the presence of inbuilt interpretation. Some classification schemes are purely morphological (even this must include some inherent interpretation), but others overtly assign genesis and/or age to mapped forms. Examples of systems incorporating the latter approaches are Klimaszewski (1961) and Demek (1972). Once interpretation becomes an integral component of the mapping scheme, the maps become a means of conveying results more than a source of raw data.

The mapping scheme proposed by Savigear (1965) (Fig. 8.1) may be used at map scales compatible with most contemporary process research and is clearly intended to be devoid of interpretation. This kind of mapping is extremely slow and tedious, and both precise identification of unit boundaries in the field and their accurate location on a base map are much more subjective than might be initially supposed. The issue of exactly how such data may be meaningfully analyzed is also very problematical. Perhaps the morphological map, like the field sketch, serves its greatest purpose in sensitizing the researcher to the subtleties of the terrain under examination.

A closely related two-dimensional simplification of morphological mapping is measurement of slope profiles. This is an honorable activity inasmuch as it has been used in innumerable contexts; indeed, there are manuals of preferred techniques (Leopold & Dunne 1971, Young 1974). Field measurement of slope profiles is extremely time-consuming and is commonly restricted to scales associated with modern process research. Slope measurements may also be derived from aerial photographs and maps, but this type of approach commonly falls into geomorphometry.

Geomorphometry

Morphometry in geomorphology is probably best termed geomorphometry simply to distinguish it from the many other spheres of morphometry (Evans 1972, p. 18). Evans went one step further and distinguished between "general geomorphometry" and "specific geomorphometry". General geomorphometry is the study of the entire landscape, while specific geomorphometry is examination of specific landforms (e.g., cirques). Pitty (1982, pp. 34–6) made a similar distinction, using "landsurface" to mean the entire topographic surface and "landform" to mean an individual component. Therefore, specific geomorphometry starts with the issue of boundaries because single units must be defined and presumably it also embraces issues of genesis, because specific landforms are invariably assigned genetic significance.

In surveying general geomorphometry Evans (1972) relied heavily on a statistical emphasis, seeking to identify those parameters that fundamentally characterize landsurfaces. He (Evans 1972, p. 22) advocated using a square matrix of altitudes at a point and derivation of measures therefrom. His basic measures were: (1) altitude; (2) slope gradient (the first vertical derivative, i.e., the rate of change of altitude); (3) downslope convexity (the second vertical derivative, i.e., the rate of change of gradient); (4) slope aspect (the first horizontal derivative); and (5) cross-slope convexity (the second horizontal derivative, i.e., the rate of change of aspect). In each instance Evans advocated calculation of the mean, standard deviation, skewness, and kurtosis of the distribution and noted the need to analyze aspect using circular statistics.

Representation of a landsurface in such a fashion is statistically sound, and certainly feasible with contemporary computer techniques and capacity. Nevertheless, a number of important geomorphic attributes are lost. As noted earlier (Ch. 6) frequency fails to express sequence, an important spatial and temporal attribute. Furthermore, if the palimpsest concept of landscape is valid, general geomorphometry may provide a very precise description of the surface at a moment in time, but how can this be interpreted genetically? In some instances (for example, strong structural control), certain landscape wavelengths may be decipherable, but in many contexts there are too many signals for any one to be interpreted with surety.

Mark (1975) took a rather more geomorphological stance than Evans in his review of geomorphometric parameters. He found horizontal components to be adequately summarized in the notions of grain (longest significant wavelength) and texture (shortest significant wavelength). Vertical scale was found to be represented by local relief (difference between highest and lowest elevations in a specified area). Mark considered that linkage of vertical and horizontal slopes was best made through mean slope; he introduced his own measure, "the roughness factor" (Mark 1975, p. 171), to link slope steepness and aspect. Finally, Mark accepted the hypsometric integral (Strahler 1952b) – see also the section on three-dimensional forms in this chapter – as the best parameter to represent altitude–area relationships.

Specific geomorphometry is subject to most of the same limitations as general geomorphometry. As an initial response it seems that boundary definition is perhaps a greater issue. This may or may not be true as many measures in general geomorphometry have been shown to be extremely sensitive to variation in the size of the areal units used for sampling. Nevertheless, specific geomorphometry would seem to offer less flexibility in boundary definition. A specific example is provided by Evans & Cox

(1974), who reported attempts to define cirques for geomorphometric analysis. While cirques have been a popular form for specific geomorphometric analyses, open rock basins, which appear to be at least as common in many glaciated areas (Sauchyn & Gardiner 1983), have not. This must be only one of many instances where only classic (i.e., fully formed or well-formed) landforms have been studied but where, in reality, the real-world distribution is much larger and includes many poorly, or partially, formed examples that have gone unnoticed.

Much geomorphometry is geometrical (i.e., involves dimensional relationships between elevation, length, etc.) but much is topological (i.e., relates numbers, e.g., the stream bifurcation ratio). Streams have been the focus of much geometrical and topological specific geomorphometry and serve as an interesting means of evaluating yet another issue in geomorphometry, that of data sources.

The bulk of geomorphometric data have been derived from topographic maps. In the U.S.A. 7.5 minute quadrangles (scale 1 : 24,000) have been a common source, although in Europe somewhat larger scales are often available. In a recent study Mark (1983) undertook a field study of an area in Kentucky that has been frequently used for geomorphometric studies. His findings with respect to first-order streams bear close scrutiny because, while he found the maps to meet U.S.G.S. mapping guidelines, Mark also found significant discrepancies between the field and cartographic representations thereof. Such research is all too infrequent but it does serve to remind us that the scale of research is an important attribute and/or constraint. Consequently, the researcher's objectives should constrain acceptable input as well as curtail extrapolation of results.

Morphological mapping requires operational definitions, but these may still be largely textual. On the other hand geomorphometry inherently requires operational definitions that are numerical. This requirement improves the likelihood of standardization during replication and is, therefore, desirable. However, while replication is improved, there is no inherent reason why geomorphometric parameters should have a strong theoretical underpinning. In examining examples of geomorphometric parameters, all stages from purely empirical to theoretically derived are apparent. A single example of a specific geomorphometric parameter from each dimensional category (linear, areal, and three-dimensional) will be considered; trends in general geomorphometry will only be considered in passing in the concluding remarks.

Linear forms

Stream form, process, and behavior have traditionally been focal to geomorphic research. One theme that has received considerable attention is that of channel plan (i.e., a bird's-eye view of a river channel). For a long time geomorphologists have used a tripartite division of channel planform with straight, meandering, and braided categories. Research embracing this scheme has primarily focused upon alluvial channels where change occurs with sufficient rapidity to permit some attempt at modeling the relevant variables; however, meanders in particular may occur in bedrock. More complex subdivisions of channel planform have been proposed (e.g., Schumm 1963a, Dury 1969), but these have failed to attain the popularity of the established tripartite classification.

Gregory & Walling (1973, p. 247) noted that meander patterns have been studied since the mid-19th century. Channel planform may be examined at a number of scales; one in which the stream may be depicted as a simple line is one of the few examples in geomorphology where attention has been directed at linear form, so it will be the focus of attention here.

Early distinctions between the three channel categories were established in purely qualitative terms. Von Engeln (1942) described meandering streams very straightforwardly and briefly, following a traditional Davisian interpretation in which meandering is an old-age phenomenon. Thornbury (1969) included early work by Leopold & Wolman (1957) and Leopold et al. (1964), but specifically noted (Thornbury 1969, p. 124) the lack of a theory to explain the dynamic behavior of meandering. Both of these textbooks reflect the long period during which channel planform classification was purely descriptive.

There are some fundamental difficulties with the standard tripartite distinctions. A primary one is that the contrast between straight and meandering channels is one of form, while the contrast between braided and the other two is one of number (Knighton 1984, p. 124). Therefore, the fundamental split should be made between single-channel and multichannel streams. This is, in fact, a much more complex issue than it would seem because it is very largely dependent upon discharge stage. Some streams, particularly proglacial ones, may even exhibit braided and meandering form as a seasonal variation (e.g., Fahnestock 1963).

Assuming that the single-channel–multichannel contrast may be established in a satisfactory manner, it is still desirable to characterize the planform of individual channels with some precision. Measures of sinuosity have been the traditional approach to this issue. The principle underlying the concept of sinuosity is that the steepest gradient between two points on the landscape is a straight line. Therefore, any circuitous route between the two points increases the distance traveled and, thereby, decreases the gradient. The ratio of the actual distance along a channel between two points on it and the straight line distance between the same points provides a measure of sinuosity.

In reality, sinuosity has not been such a simple concept in geomorphology. This is because the straight-line distance has been subject to a

Table 8.1 A selection of sinuosity measures to show its actual variability.

Measure	Source
$\dfrac{\text{length of a channel in a given curve}}{\text{wavelength of the curve}}$	Langbein & Leopold (1966)
$\dfrac{\text{thalweg length}}{\text{valley length}}$	Leopold *et al.* (1960)
$\dfrac{\text{stream length}}{\text{valley length}}$	Schumm (1963a)
$\dfrac{\text{channel length}}{\text{length of meander belt axis}}$	Brice (1964)

number of differing definitions (Table 8.1). Leopold *et al.* (1964, p. 296) used a sinuosity value of 1.5 to define meandering. A similar approach by Brice (1964) used a sinuosity index (*SI*) with a slightly different definition. Brice then categorized streams as straight (*SI* < 1.05), sinuous (*SI* = 1.05 to 1.5), and meandering (*SI* > 1.5). As noted by Leopold *et al.* (1964, p. 296) sinuosity is by no means a perfect measure of meandering because most streams are sinuous to some degree, and the concept of meandering usually embraces the notion of rhythmical or symmetrical changes of direction, a characteristic that is not captured by sinuosity.

Richards (1982, p. 182) provided a serious objection to the entire process of categorizing stream channel planform by sinuosity. He regarded classification based on arbitrary sinuosity values as an entirely illogical procedure because there is a continuum of channel patterns that may be related to stream power. By using total sinuosity (Le Ba Hong & Davies 1979), which is the ratio of the totally active channel length (i.e., in the case of multiple channels, the length of all of them is summed) to valley length, it is possible to encompass both single-channel and multichannel streams by a single criterion.

Creation of sinuosity indices really marked the beginning of a transition from purely qualitative definitions of channel plan to the linkage of form and theory. An index itself is merely a quantitative description; as such it is superior to a qualitative description (primarily because it helps in the process of repetition or replication), but is not intrinsically a theoretical step.

Leopold & Wolman (1957) introduced the concept that channel pattern is controlled by the interaction among a set of variables. They also argued that as a result of this interaction there should be a continuum of channel forms and that straight, meandering, and braided channels should intergrade. The underlying geomorphic argument will not be discussed here. Schumm published a number of papers (of which the most important is probably the Schumm (1963a) paper) in which he related channel cross section, river transport load, and channel planform to each other (see Miller & Onesti (1979) for an interesting commentary on the statistical techniques used by Schumm). Most contemporary textbooks on fluvial geomorphology (e.g., Knighton 1984, Morisawa 1985, Richards 1982) contain comprehensive discussions of the topic.

As mentioned earlier, meander geometry is not fully embraced by the concept of sinuosity because the symmetry of the bends is not characterized. Symmetry may actually be modeled by depicting meanders as sine-generated curves. Such an approach was undertaken by Langbein & Leopold (1966). Their work was deeply rooted in a theoretical approach because they based their use of sine-generated curves on the idea that such curves represent the most probable distribution statistically because they incorporate the fewest changes in direction and, therefore, represent the least work. The concept of equalization of energy expenditure is involved here as it is believed that more energy is dissipated by friction at a bend than is dissipated over a straight reach. Yang (e.g., Yang 1971, Yang & Song 1979) has pursued the concept of the minimization of the rate of energy dissipation at length.

Richards (1982, p. 197) pointed out that sine-generated curves cannot model the up-valley segments included in acute meander bends; while Ferguson (1973, 1976) discussed several other models for meander bends. In fact, meander modeling has now progressed to a number of other techniques including series analysis using either serial correlation or spectral analysis.

While meander planform is now firmly wedded to hydraulic and hydrologic theory, identification of precise relationships between them is unlikely to emerge soon. Thornes (1979) commented upon the fundamental split in fluvial modeling between deterministic approaches (favored by J. H. Mackin) and empirical work such as that published by L. B. Leopold, W. B. Langbein, T. Maddock Jr, and M. G. Wolman (to name just a few of the better-known early practitioners of the approach). He went on to suggest that it is the fact that no branch of physics has focused upon two-phase (fluid–solid) flow that has left geomorphology bereft of a suitable platform from which to launch deterministic stream modeling. If the empirical, probabilistic path is followed, as has generally been the case, then the interaction between variables and the availability of several possible responses as adjustments to any input must be emphasized. As Maddock (1970, p. 2309) wrote, "The behaviour of alluvial channels is a matter of tendencies."

A final source of difficulty is to appreciate that neither hydraulic nor hydrologic theory is likely to provide a complete answer to meander patterns. The most obvious difficulties to be encountered stem from the timescales over which geomorphologists are interested and from spatial constraints and interference created by valley-side topography. Temporal issues are well-illustrated by Dury's (1964, 1965) research on underfit streams; while both the spatial and deterministic/probabilistic issues are embraced in Ferguson's (1976) meander modeling.

It is quite apparent that meanders represent a linear form whose investigation has undergone many stages of qualitative and quantitative evolution. As an example meanders have the advantage of being a topic that has received prolonged and diverse analysis; conversely, they exhibit the disadvantage that theoretical approaches have yet to clarify the situation in a universally accepted fashion. The situation is still not a clear one, beyond general recognition of a relationship between stream discharge and meander wavelength (and associated measures).

Meandering also provides a good case study for consideration of equifinality as a valid concept in geomorphology. This is because meandering occurs not only in streams, where sediment load is usually considered a contributing factor, but also in features such as the Gulf Stream, supraglacial streams, and jet streams. In the latter three cases temperature gradients are probably important contributing factors, but sediment load is certainly not.

Obviously, only a very small portion of the available material has been selected to illustrate some salient issues relating a linear form to process(es). Anyone interested in pursuing the topic of meandering in detail would do well to start with the reviews provided by Knighton (1984), Morisawa (1985), and Richards (1982).

Areal forms

A fluvial theme may readily be continued when examination of form is extended to two-dimensional, or areal, contexts. The shape of drainage basins has long been studied and subjected to a number of different measures and will be the focus of attention here. Prior to examining the various measures designed to depict drainage-basin shape it is worth emphasizing that in all instances it is only the two-dimensional outline that has been considered and no attempt to incorporate an elevational or topographic component has been made.

Horton (1932) wrote a paper reviewing a sizable array of measures that had been proposed to capture a number of drainage-basin characteristics. Two measures of basin shape were mentioned by Horton (1932, p. 351), the form factor and compactness. The form factor (F) is the ratio between basin width and length, the latter being measured from the stream mouth to a point on the basin perimeter opposite the head (mouth) of the main stream. Horton formalized this measure as

$$F = M/L^2$$

where M is the basin area in square miles and L is its length in miles. Compactness (C), a measure that Horton attributed to Gravelius, is the ratio of the perimeter of the drainage basin to that of a circle of equal area. He proposed calculating this value as

$$C = \text{perimeter}/2\sqrt{(\pi M)}$$

(where M is again basin area). Note that C has a minimum value of 1, which occurs with a circle.

Horton's paper focused on hydrology and hydrologic responses; accordingly, he appraised the two measures for their predictive capacity on these fronts. He found the form factor to have some utility in predicting maximum flood discharge in elongated basins, but little utility in irregularly shaped basins or those with permeable soils. He found compactness to serve little purpose at all.

These appraisals by Horton raise the issue of why drainage-basin shape should be studied. Clearly, Horton's viewpoint focused upon one substantive issue, namely, the relationship between basin shape and hydrologic behavior of the enclosed stream(s). It is not difficult to imagine that one might readily hypothesize relationships between basin shape and/or size and the manner in which this would influence runoff, throughflow, and groundwater, thereby controlling the hydrologic response of the stream(s) within the basin. Another obvious realm of interest would be to consider the interaction between fluvial erosion and landscape morphology. This topic might be viewed temporally – the evolution of basin shape through time – or spatially – for example, basin-shaped variability with changes in size, rock type, or climate. Any such modeling might be undertaken empirically or deterministically. Ratios such as those mentioned so far are simply indices that aid replication but lack any theoretical underpinning. Other measures have been used and may also be evaluated with respect to their origin or underpinning, as well as their usefulness.

The United States Army, Corps of Engineers (1949) modified Horton's form factor by inverting it, thus

$$F = L^2/A$$

where L is basin length and A is basin area. This appears to offer no substantive improvement over Horton's calculation, although obviously it produces different numerical values.

Both the above form factors are dimensionless, as they involve dividing length squared by the same dimension (see Ch. 12). Miller (1953) also proposed a dimensionless ratio called the circularity ratio (C). This was defined as

$$C = A_b/A_c$$

where A_b is drainage-basin area and A_c is the area of a circle having the same perimeter. Schumm (1956) proposed an elongation ratio (E) defined as

$$E = d/L_m$$

where d is the diameter of a circle with the same area as the drainage basin, and L_m is the maximum length of the basin parallel to the principal drainage line. Again both of these parameters are dimensionless measures. An approach retaining comparison with known geometric forms, but not exclusively circles, was proposed by Lee & Sallee (1970) but appears to offer no obvious advantage.

Development of indices founded on dimensionless ratios and circularity incorporate two theoretical concepts. A dimensionless approach was believed during the 1950s to offer an avenue of geomorphic research that would preclude scale as a source of variability. A classical development of this particular theme appeared in Schumm (1956) with comparison of natural badlands in Arizona and South Dakota with those developed on the flanks of an abandoned clay pit in New Jersey. Circularity is an important theoretical yardstick because a circle encloses the largest area for a given perimeter of any geometric form. This means that a circle may be envisaged as the lowest free-energy form, i.e., the form that is least vulnerable to external influences. Such a situation is important because nature is widely considered to tend toward lowest-energy states. Any index founded on some measure of circularity is, thus, based on this minimum free-energy concept. Circularity and approximation of spherical form were incorporated in a ratio proposed by Krumbein & Graybill (1965). They suggested using the ratio of the diameters of the inscribed circle (i.e., the largest circle that may be drawn within the basin) and the circumscribed circle (i.e., the smallest circle that may be drawn to enclose the basin). This particular measure was clearly inspired by a two-dimensional approximation for sphericity that has been used in microscopic studies of sedimentary particles (e.g., Riley 1941).

Consideration of an individual drainage basin would not lead one to expect it, intuitively, to be circular because a stream is a distinctly elongated form. Conversely, considered from the perspective of optimal packing, drainage basins might be expected to be hexagonal (Woldenberg 1972) (see Ch. 7). Chorley et al. (1957) pursued the theme of individual drainage-basin shape by recognizing the inapplicability of circular form and proposing one loop of a lemniscate (Figs. 8.2a and b) as an appropriate yardstick. Qualitatively, a lemniscate may be described as a petal shape, tear shape, or pear shape; in this instance its long axis (l) is analogous to drainage-basin

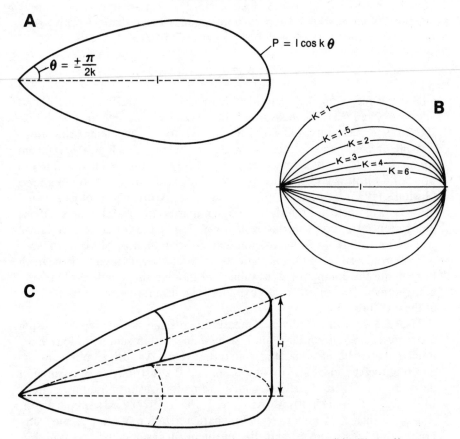

Figure 8.2 The lemniscate. (a) Defining parameters. (b) The effect on lemniscate shape of varying k. (c) The three-dimensional projection of the lemniscate used to create a measure comparable to Strahler's hypsometric integral. The line intersecting l divides the area of the figure into half. (From Chorley & Morley (1959).)

length. Mathematically, a lemniscate is described in radial coordinates by the formula

$$p = l \cos (k\theta)$$

where θ lies between $-\pi/2k$ and $+\pi/2k$, l is the longest diameter of the loop ($p = l$ when $\theta = 0$), and $k \geq 1$ is a constant (when $k = 1$ the loop is a circle).

The shape of a lemniscate loop is extremely sensitive to variations in k (Chorley *et al.* 1957, p. 139) as

$$k = l^2\pi/4A$$

where l is as defined above and A is loop area (Fig. 8.2b). However, while the value of k gives a good general impression of basin shape, Chorley *et al.* (1957, pp. 139–40) developed a lemniscate ratio (K), calculated as

$$K = P/P_m$$

where P is the perimeter of an ideal lemniscate with the basin's calculated value of k, and P_m is the measured value of the basin perimeter. Obviously, the closer K is to unity, the closer the basin is to a perfect lemniscate loop.

While it is easy to see, intuitively, that a circle is not a logical ideal form for an entity driven by a linear feature such as a stream, the lemniscate did not fair well in a comparative test in which Morisawa (1958) compared Horton's (1932) form factor, the United States Army, Corps of Engineers' (1949) form factor, Miller's (1953) circularity ratio, Schumm's (1956) elongation ratio, and Chorley *et al.*'s (1957) k value for their correlation with basin runoff. In order to eliminate scale influence, Morisawa (1958, p. 590) used a runoff : rainfall ratio as an independent variable with which to evaluate the comparative usefulness of the five shape indices. Only the elongation and circularity ratios exhibited significant regression coefficients at the 0.05 probability level.

The ease with which individual measurements may now be repeated using a computer and the intense interest in shape within sedimentology have impinged upon basin-shape studies in recent years. An early example, incorporating the center of gravity of a basin and the distribution of basin area derived from digitized perimeter measurements, was proposed by Blair & Bliss (1967). More recently McArthur & Ehrlich (1977) used a simple extension of Ehrlich & Weinberg's (1970) characterization of grain shape using Fourier analysis to investigate the problem of basin shape. As noted in Ehrlich & Weinberg, one inherent problem in describing shape is identification of a variable that unambiguously distinguishes shape characteristics of different scales (e.g., a grossly circular shape and a smooth or extremely irregular surface are not conflicting concepts, but complementary).

These problems may be addressed in a consistent and reasonably objective manner using Fourier analysis. First, it is necessary to establish a coordinate system; second, the basin's (or any other form's) center of gravity must be determined. In Fourier analysis the radius of the periphery is then expanded as a function of the angle about the basin's (form's) center of gravity $R(\theta)$. The equation has the form

$$R(\theta) = R_0 + \sum_{r=1}^{\infty} R_r \cos{(r\theta - \phi_r)}$$

where R_0 is the average basin radius, R is harmonic amplitude, r is harmonic order, ϕ is phase angle, and θ is polar angle (in this case measured from the basin mouth). It is possible to envisage such a form as follows: the zero-order harmonic is a simple circle with an area equal to that of the basin;

the first harmonic is another circle, but it is offset; the second harmonic is a figure eight; the third harmonic is a trefoil (Ehrlich & Weinberg 1970, p. 206). Phrased rather more colloquially, there are as many "bumps" superimposed on the initial form as the order of the harmonic. Therefore, as the number of harmonics is increased, the degree of detail also increases. There must be at least twice as many data points measured on the periphery of the form as there are harmonics used in the analysis; 10 harmonics are commonly used to describe basins or grains.

Harmonic analyses create two variables in shape equations, harmonic amplitude and phase angle. These may be displayed in graphic form and two or more such plots may be analyzed (compared or contrasted) using a variety of standard statistical tests. Such an approach is now a common method of evaluating grain shapes in sedimentology, but has not yet been widely used in basin-shape studies. A much simpler version of this kind of approach was used by Boyce & Clark (1964), who measured the distances of radials (at a regular angle interval) drawn from the basin mouth to its perimeter. Their index was then calculated as (Boyce & Clark 1964, p. 568)

$$\sum_{i=1}^{n} \left(\frac{r_i}{\sum_{i=1}^{n} r_i} \times 100 - \frac{100}{n} \right)$$

or may be expressed as (Gardiner 1976, p. 25)

$$BC = \sum_{i=1}^{n} \left(N \frac{R_i}{\sum_{i=1}^{n} R_i} \right) \times 100 - \frac{100}{n}$$

when n (or N) is the number of lines drawn as radials from the basin mouth at equal angle intervals and r (or R) are the lengths of the radials. Annotation has been expanded slightly in both equations to increase clarity.

McArthur & Ehrlich's (1977) study was a comparative one in which Fourier analysis was evaluated for its effectiveness against some established basin-shape ratios, namely form factor (Horton 1932), circularity ratio (Miller 1953), elongation ratio (Schumm 1956), lemniscate index (k) and lemniscate ratio (K) (Chorley et al. 1957). McArthur & Ehrlich (1977, p. 294) found the circularity and elongation ratios to be sensitive to large-scale regional contrasts, but not intermediate-scale variability; and they found the two lemniscate values to discriminate basin-shape variability poorly. Not surprisingly the Fourier analysis, which inherently contains more information within it, was the best discriminator of basin-shape variability in their opinion.

A rather different approach to basin shape was pursued by Ongley (1970). His approach may be described as a dynamic and weighted one, but its primary distinction from those mentioned to this point is that it is

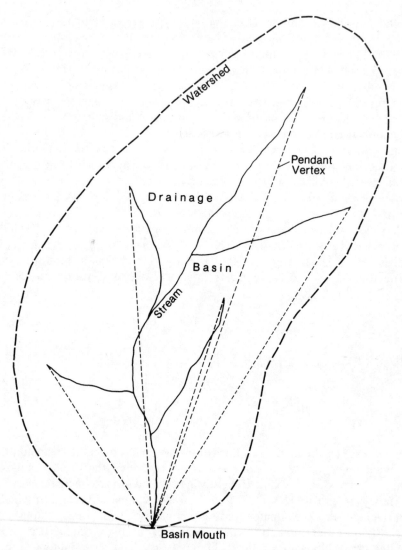

Figure 8.3 An illustration of Ongley's pendant vertices. (From Ongley (1970).)

founded upon the stream network within the basin (Fig. 8.3). The source point of each stream within the basin is joined to the mouth of the basin by a straight line ("pendant vertex"; Ongley 1970, p. 39). The azimuth and length of each line, the latter being "vectorial equivalents", are then measured. The distribution of angles and lengths may now be treated as a population frequency distribution and the moments (mean, standard deviation, skewness, and kurtosis) calculated. Ongley commented on the difficulty of analyzing the circular distribution created by the azimuth data, but as noted in a previous chapter it is perfectly possible to treat it with appropriate circular statistics (Mardia 1972), or by less appropriate linear approximations. First-order basins are a problematic issue as they contain only a single stream, but Ongley (1970, p. 39) accepted another measure as an alternative in such instances.

It is apparent that none of the measures designed to monitor drainage-basin shape has any well-founded theoretical foundation beyond the notion that "all things in nature tend to the lowest free-energy state". In general, the measures discussed have been related to stream discharge rather than to any geomorphic characteristics. Even the hydrologic link is a purely empirical one, with occasional success in establishing a fuzzy relationship between basin shape and stream hydrograph. Indeed it is difficult to see how something as complex as a drainage basin would ever exhibit widespread uniformity between a single form measure and the fluvial and mass wasting processes that influence it. Basin shape, for example, is probably dominated by mass wasting rather than fluvial processes, and this will certainly be the case where a floodplain serves to buffer the bulk of the basin from river behavior.

The fundamental issue at hand in drainage-basin shape is that there is no theoretical thread linking basin shape to either hydrologic or geomorphic processes. Without this link any index is of questionable value. While it may be possible to compare the behavior of various measures over the range of selected samples, the only conclusion that may be drawn is that some are highly variable (sensitive) while others are relatively stable (insensitive). Even in the case of sensitive measures the question "sensitive to what?" remains unanswered.

This evaluation is not intended to single out the research discussed for particularly harsh criticism (much of it has been highly innovative), but it does serve to highlight the extreme difficulty of relating process to external areal form, particularly simple measures thereof. Furthermore, it reveals the futility of examining form alone and then trying to reason backward to genesis. As noted in the previous chapter, and emphasized by Getis & Boots (1978), form comparisons serve science best (only?) when used to test a previously established hypothesis.

Three-dimensional forms

Whatever the scale of their interest geomorphologists are ultimately interested in three-dimensional landforms. Investigation of three-dimensional form is inherently difficult; hence, the widespread use of two-dimensional substitutes (e.g., the slope profile, the channel cross section, or the longitudinal profile of a river). In the examples cited in the preceding sentence, change along the third dimension is simply ignored, but in the case of the hypsometric integral an attempt is made to collapse three

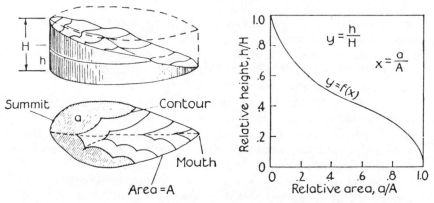

Figure 8.4   The salient features of the hypsometric integral and its calculation. Here H is the vertical height between the mouth of the basin (i.e., the stream outlet or junction) and the highest point on the watershed; h is the height of the contour under examination above the mouth of the basin; A is the area of the entire basin as defined by the watershed; and a is the area of the basin that is above the contour h. By expressing height and area as h/H and a/A, each value becomes a relative one with a maximum value of 1 and a minimum value of 0. Individual values of h/H and a/A may then be plotted against each other to produce the line (called the hypsometric curve) in the right-hand part of the figure; the hypsometric curve always runs from top left to bottom right, but its shape varies. The hypsometric integral is the area under the hypsometric curve (i.e., the land mass still to be eroded) expressed as a percentage of the entire rectangle. (From Leopold *et al.* (1964), after Strahler (1952a).)

dimensions into a two-dimensional framework. The hypsometric integral will serve as the three-dimensional case study as it not only represents an attempt to wrestle with a difficult issue, but also sustains the fluvial theme pursued in the preceding two sections.

The hypsometric integral is an attempt to create a measure of the distribution of land mass within a drainage basin. In geomorphology a paper by Strahler (1952b) was the real starting point of the concept, although Strahler himself noted its earlier use in hydrological studies. A substantial portion of Strahler's paper was devoted to a Davisian (Ch. 10) interpretation of results, as well as suggested modifications thereto; this geomorphic interpretation of the hypsometric integral will be ignored here and attention restricted to creation and morphological meaning of the integral.

The mechanics of producing a hypsometric integral are fairly time-consuming; they are described fully in Figure 8.4. The essential steps are: (1) define the boundary of the drainage basin and then measure the area of the basin; (2) measure the area between each pair of contours within the basin; (3) identify the elevation of the highest point in the basin and the elevation of the stream mouth. Results of these steps may be plotted on a graph by displaying elevation on the vertical axis, and area above each elevation on the horizontal axis. Displayed in absolute units such a plot is called a hypsographic curve.

An important theme in much of Strahler's pioneer research into quantitative geomorphology was identification of scale-free principles underpinning geomorphology. This concern is exemplified in his development of the hypsometric integral. He refined the hypsographic curve by expressing it in relative terms instead of absolute terms. The height of a contour above the mouth of the stream is expressed as a proportion of the total height of the basin, and the area is expressed as a proportion of total basin area (Fig. 8.4). Both vertical and horizontal axes then become restricted to a range from 0 to 1. Furthermore, the graph may then be conceptualized as a rectangle (Fig. 8.4). The rectangle so depicted represents the amount of material that would have been included within the area of the present basin had the basin been eroded in an initially horizontal surface (clearly, a very questionable assumption, and one for which there is very rarely any evidence).

Defined in this fashion, the hypsometric curve will always run from top left to bottom right; but the shape of the curve will reflect the distribution of land mass with elevation within the basin. The hypsometric integral is then defined as the ratio of the area within the rectangle beneath the hypsometric curve to that of the entire rectangle. Expressed as a percentage, it may be taken to represent the amount of material still available for erosion.

Strahler (1952b, pp. 1121–5) then developed a function to describe the hypsometric curve mathematically. The equation developed by Strahler focused attention upon the inflection point between the concave and convex segments of the hypsometric curve; this has physical significance because it marks the elevation "at which the rate of decrease of mass upwards changes from an increasingly rapid rate of decrease to a diminishing rate of decrease" (Strahler 1952b, p. 1123). The second value receiving attention is the exponent in the equation; when its value is less than 1 hypsometric curves are s-shaped, but when its value is greater than 1 then curves are simple concave forms. The examples in Strahler's paper include use of these values as interpretive tools; however, Strahler's extensions and developments beyond the hypsometric integral itself have never been widely used.

Chorley & Morley (1959) extended the use of lemniscate (Chorley et al. 1957) to a three-dimensional approximation (Fig. 8.2c) of the hypsometric integral. Computational steps are quite complex and the parameter has never been widely used. Pike & Wilson (1971) demonstrated that the elevation–relief ratio (E), defined by Wood & Snell (1960), is actually mathematically analogous to the hypsometric integral. E is calculated as

$$E = \frac{\text{mean elevation} - \text{minimum elevation}}{\text{maximum elevation} - \text{minimum elevation}}$$

(all values being from with the basin under consideration). The advantage of E is that mean elevation is determined by a point-sample technique, and this bypasses the time-consuming task of measuring between-contour elevations using a planimeter – the traditional objection to use of the hypsometric integral.

Gardiner (1974), in an extensive review of drainage-basin morphometry, was unable to do more than note the general superiority of the hypsometric integral, while also noting the limited morphological information encompassed in any of the proposed measures. Three-dimensional parameters of basin morphology have served some useful purposes in predicting hydrologic behavior, but have yet to achieve a substantive role in geomorphic research. Part of this limited success stems from the difficulty of expressing three-dimensional form succinctly in a numerical form that readily conveys meaning or creates a mental image of shape. Another portion of the difficulty emanates from a lack of theoretical underpinning: shape is being described (more or less, depending upon the parameter), but there are no obvious functional interpretations of the depicted forms. Until there is some sort of theoretically defined expectation for form, any index or indices will be of very limited usefulness.

Conclusions

Form has always been *the* focal interest in geomorphology; while scales of interest, methodologies, and techniques may all have changed, there has been little or no deviation from the original geomorphic goal – namely, explanation of land*form*. If a truly simplistic viewpoint is taken, geomorphic processes may be called causes and landforms may be called effects. Of course, Schumm & Lichty (1965) showed this to be a time-dependent set of relationships, and systems modeling (Ch. 12) has revealed the importance of feedback relationships. Ignoring these complications momentarily, it is easy to see why dissatisfaction with early, qualitative descriptions of form would lead those seeking to improve the situation toward investigation of causes.

Dissatisfaction with Davisian geomorphology (Ch. 10), the dominant approach in the English-speaking world prior to about 1950, produced several changes in geomorphic research. Davis's benign neglect of process undoubtedly did much to direct an innovative generation of geomorphologists toward process research. In turn, this resulted in a profound reduction in the scale at which research was conducted because it was, and remains, infeasible to conduct process research at a regional scale, save through a statistical sampling strategy. However, it is worth emphasizing that the initial upsurge in process geomorphology (of which R. E. Horton and A. N. Strahler may be labeled early leaders) combined process measurements with overt attempts to produce numerical form descriptions.

It is apparent that during the intervening years process studies have burgeoned while form studies have withered. Field studies have generally focused upon process rates, while laboratory research has increasingly focused upon the fundamental chemical and physical nature of geomorphic processes. Finally, on the theoretical front, progress has been made on both the temporal and spatial behavior of processes, although temporal behavior appears to have received more attention and our grasp of it appears to be presently the more sophisticated of the two.

During this same period geomorphometric studies have slipped out of the mainstream of geomorphology in the English-speaking world. Contemporary computers provide those interested in general geomorphometry with the necessary capacity and speed with which to process regional studies. In the U.S.A., the U.S. Geological Survey (McEwen *et al.* 1983) has been committed to producing digital elevation models (D.E.M.) compatible with 7.5 minute quadrangles, while the Defense Mapping Agency has been pursuing a similar approach by digitizing the 1:250,000 map series. These efforts have produced good-quality, standardized data bases that have stimulated considerable research interest in cartography and remote sensing. However, there has been relatively little interest shown by geomorphologists. When such work has been undertaken, it has invariably fallen under the rubric of terrain analysis. Terrain analysis has been strongly influenced by military needs (e.g., tank trafficability), but appears (with notable exceptions) to have contributed very little to theoretical geomorphology.

Craig (1982a, b) provided one example of the utility of what was essentially a D.E.M. to investigate ergodicity (although Church (1983) raised some questions about the technique). Using a section of the Appalachians, Craig found landforms there to oscillate about, but never attain, stability. While such findings are clearly of great theoretical importance, most contemporary computer models of hillslope development, be they two- or three-dimensional, have been founded on numerical modeling of processes, rather than of geomorphometry.

One alternative mathematical approach to founding topographic models on elevation matrices is to use an analytical approach (Eyton 1974). This means that instead of having height values in a matrix as the basic building blocks of the model (as in a D.E.M.), the researcher uses calculus to develop equations that describe continuous surfaces ("surfacing" being the technical term used to describe the technique in which such numerical models are created). Information on individual points is then obtained by solving the equation for that point. Derivation of functions (e.g., the first two derivatives are slope in the x and y directions) from analytical models is a flexible and powerful approach. However, development of analytical models that describe complex landform assemblages adequately for the purposes of general geomorphometry is very complicated and time-consuming. Nevertheless, an analytical approach may be used successfully over the more restricted spatial scales commonly encountered in specific geomorphometry. Hansel (1980) provided an interesting example of both specific geomorphometry and an analytical technique to differentiate the genesis of sinkholes.

A number of issues serve to confound use of geomorphometry. Among the most important is the need to identify situations where a landform is in equilibrium (whatever form that may take) with prevailing processes. This goal may be unattainable in the field, if the palimpsest concept is of widespread applicability. Therefore, an appealing alternative approach would seem to be numerical modeling. Another fundamental issue is that of equifinality, whether defined traditionally or in the manner preferred by Haines-Young & Petch (1983). Defined traditionally, equifinality robs geomorphologists of the ability to assign unique genesis by form alone; defined in the manner suggested by Haines-Young & Petch, landforms in balance with a process become decipherable, but only with respect to the prevailing process; those in imbalance are indecipherable.

For those interested in specific geomorphometry, operational definition of boundaries is a vital, early step. Intuitively, it would seem to be a relatively easy one, but experience shows it to be extremely complex, and the decisions embraced by it profoundly important. In a tangential sense, the internal properties of landforms represent the same kind of problem. A good example of this would appear to be the extensive literature on the specific geomorphometry of drumlins. While copious, it seems that this literature has failed to contribute nearly as much to understanding of drumlins as the stratigraphic work pursued by D. M. Mickelson and others (e.g., Whittecar & Mickelson 1979, Stanford 1983) at the University of Wisconsin at Madison.

Form can be an important medium for fundamental research if, as Getis & Boots (1978) have emphasized, it is preceded by rigorous hypothesis formulation. It would seem that geomorphologists have now advanced their knowledge of processes to the point where such hypothesis formulation is possible and, consequently, specific geomorphometry is potentially a profitable venture. Progress will most probably result from work conducted on numerical models where the bewildering complexities of the real world or field are replaced by the simplicity of abstraction. It seems unlikely that present understanding of regional geomorphic processes is adequate to support extensive theoretical development of general geomorphometry. However, it is geomorphometry (implying numerical precision), wedded to hypotheses founded on process geomorphology, that affords the opportunity, not qualitative techniques nor indices or parameters bereft of theoretical underpinning. Those unconvinced of the opportunities presented by rigorous examination of form should consider the contribution and durability of D'Arcy Wentworth Thompson's *On growth and form* (the abridged edition will suffice (Thompson 1961)), and in the contemporary scene the enormous impact that Mandelbrot's (1982) treatise on fractal geometry appears likely to have on our very perception of form in the natural sciences.

9 The fundamentals – a summary

The purpose of this chapter is not to summarize preceding chapters in detail but to reinforce the concepts contained in them by emphasizing the truly fundamental issues confronting geomorphologists. Attempting to distill the quintessential from the essential is in some ways akin to alchemy, innately appealing, but laced with pitfalls. However, pitfalls notwithstanding, the broad intellectual issues must be faced because they represent challenges that can never be overcome by technical improvement alone.

Geomorphologists are concerned with explaining the shape(s) of portions of the Earth's surface. Evidence at hand indicates that the great majority of these features develop over very long spans of time as measured against a human lifespan. Clearly there are exceptions, particularly as spatial scale decreases, but the essential validity of the statement is unassailable. The magnitude of this problem increases as the geomorphologist's goals become more sophisticated, because the more detailed the data requirements the shorter is the effective record.

This profound constraint immediately forces geomorphologists to seek analogs. Obviously, the only sources are the present and historical (i.e., written) records, both of which are extremely brief by geomorphic standards. This situation would seem to elevate the conceptual structures used to extrapolate very short records over much longer timespans to a preeminent position in geomorphology.

Logically, this situation places a tremendous premium in the first instance on understanding present-day (spatial) geomorphology and the justification of the traditional aphorism "the present is the key to the past" emerges. However, this concept is only a solid starting point if the geomorphologist believes that present land shapes exhibit "characteristic forms" with respect to present-day processes. In turn, if this stance or posture is taken, retrodiction and prediction become extremely limited because, by definition, present forms will lack imprints of their past and obviously cannot have "memories" of their futures.

The alternative line of reasoning to that of characteristic-form modeling is to assume that landforms exhibit relaxation forms. This posture casts great doubt on the worth of present-day process–form studies because not only may present processes not have produced the forms observed, but present forms may well condition present processes.

Characteristic-form and relaxation-form models are founded on inherently conflicting assumptions. Whichever viewpoint the geomorphologist takes there will always be contradictory evidence available; in fact, the researcher is really being asked to identify the starting point on a circle. This situation would seem to inspire two questions: (1) Is there a definitive or generic way out of the dilemma? (2) Is there any way out at all? I believe the answer to the first question is "No" (thereby ending my discussion of it!), but that the answer to the second question is "Yes".

The solution to the problem posed in the second question is to have a clearly defined research objective and a well-developed grasp of relevant theory. If indeed the research geomorphologist has a full grasp of relevant theory he/she will appreciate not only its strengths but also its inevitable limitations and weaknesses. In effect, he/she has to recognize that an artificial entry point into a circular argument has been made. Furthermore, the geomorphologist will inexorably be drawn into undertaking research that contributes to theory because he/she will not be able to ignore that theory is pervasive. This reality may be underlined by paraphrasing an earlier quotation from Goodman (1967) (See Ch. 2) – a fact is a small theory and a theory a big fact.

Greater emphasis upon theory would also appear to lead, axiomatically, to acceptance of the notion of corroboration rather than that of verification. In general, Popper's critical rationalism represents the most satisfactory intellectual perspective, although there is no denying that Kuhn presented a valid picture of the way the scientific community tends to operate. There is a clear distinction between healthy skepticism and intellectual paranoia, and the former needs to be encouraged. In general terms, the more widespread are theoretical concerns among practicing geomorphologists, the more robust the discipline's foundation is likely to become. In much the same way as democracy needs an active electorate to select and police elected officials, an academic discipline needs universal concern with theory to insure intellectual vigor.

The details of acceptable scientific procedures in geomorphology have already been outlined and will not be reiterated here. However, it is worth re-emphasizing one combination of general concepts. Popper's restriction of the measures of satisfactory scientific procedure to formal presentations seems an eminently useful one. The worst kind of legislation is that which is unenforceable or unenforced. This principle is clearly applicable to scientific procedure. It is very apparent that creativity is a process of which we have an extremely poor understanding. Therefore, not only is no purpose served by declaring unenforceable ways in which it should be produced, but active harm results from suggesting that some ways are unacceptable. On the other hand it is equally desirable to have a set of rules by which quality control is ensured; Popper's approach would seem to embrace both these principles.

The Popperian contribution to quality control in science is concrete while that to scientific creativity is essentially no more than a hands-off view. Given that creativity is a mystical process and at its pinnacle is probably truly incomprehensible, it may still be fostered in a general sense. Johnson's (Chamberlin's) multiple-working-hypothesis approach certainly serves creativity as well as such an elusive activity can be served by routine guidelines. Consequently, the approaches advocated by Popper and Johnson (Chamberlin) merit attention as a fundamental pairing in scientific procedure.

Turning attention to the content of geomorphology it is apparent that geomorphologists are inherently interested in both erosion and deposition. However, it is quite obvious that there is no direct erosional record because by its very nature erosion removes the evidence itself. Technically, the age of the remaining, uneroded bedrock provides an upper limit on the duration of erosion, but in practice this is invariably of very little help. This set of circumstances places tremendous emphasis on understanding the nature of the stratigraphic record. In fact, this record is not only the sole long-term one for geomorphologists, but also for most other Earth scientists, including climatologists.

Ager (1981) produced a very stimulating and provocative review of stratigraphy. One of Ager's (1981, p. 35) points, that of the "phenomenon of the gap being more important than the record" deserves particular mention here. In emphasizing that the stratigraphic record is nowhere near complete, Ager presents a considerable challenge to geomorphologists as well as stratigraphers. Quite simply, neither group can ever expect to know the geomorphic/geologic past in its entirety. Even to interpret a complete depositional record in terms of both erosion and deposition would require a powerful theoretical framework. However, to undertake the task utilizing only a highly fragmented depositional record greatly increases the need for a very refined theoretical foundation.

Another salient aspect of stratigraphy is that its practitioners must turn to sedimentologists and geomorphologists for present-day analogs so that they may interpret their data. Without such analogs stratigraphy cannot exist. However, while the proper interpretation of the stratigraphic record may tell of the nature of paleo-environments and paleo-events, it provides only the very crudest kind of relative dating (e.g., such basic concepts as older deposits laying beneath younger ones).

Precise, absolute dating necessitates use of isotope techniques (which together with many others have been reviewed by Bradley (1985)). Such work is important for geomorphologists because of their innate interest in process rates, as well the nature of geomorphic processes. It is important to remember that dissatisfaction with Davisian geomorphology was in large part initiated by improvements in dating techniques. While geotechnical studies may improve geomorphologists' understanding of the nature of geomorphic processes, like a chemical equation, such studies tell only what is possible and not the rate at which it occurs in nature (which surely varies both spatially and temporally). Consequently, geomorphologists will inevitably have their temporal, and thus their process-rate, concepts, dictated by isotope studies. Vita-Finzi (1973) emphasized this reality, and while the specific isotopes may change there seems little likelihood that the general principle will.

If geomorphologists must wrestle with complex temporal issues, so too must they confront equally complex spatial ones. Perhaps no purely spatial issue is as focal as scale linkage. The inability to undertake precise field measurements on a grand scale is a severe limitation in a discipline that attempts to embrace the entire surface of the Earth (and increasingly even other planets). Getis & Boots (1978) emphasized that the study of pattern alone could never provide a causal explanation, but that it provided one means of testing previously articulated hypotheses. In an analogous fashion it is possible to argue that study of form alone in geomorphology will experience the same fate. If this is the case, hypothesis generation is likely to derive, at least in part, from process research. In turn, process research is undertaken at extremely restricted scales for a variety of fiscal and intellectual reasons. Therefore, geomorphologists are once again thrown back into a position where the intellectual structures they use to extrapolate results from one scale to another are focal. If they are unable to project their process concepts from the scale at which they are undertaken to larger scales, they are axiomatically precluded from testing any concepts about larger-scale forms. The final twist in the geomorphic knot is the infinite variety of ways in which space and time interact. Comprehensive resolution of this problem is simply too difficult a task at this time and will not be attempted here. Instead the points made so far in this chapter will be reinforced by identification of two very broad strategies that geomorphologists should use more widely.

First, Nystuen (1963) pursued abstraction, arguing, as have many others, that the light shed by simplicity is greater than that found while trying to unravel the complexities of reality. This may well take the form of numerical modeling in geomorphology, but it need not be pursued exclusively in the language of mathematics. Second, Church (1984, pp. 563–4) has presented a case for true field experiments in geomorphology. In forwarding this suggestion he pointed out that such an approach requires: (1) a conceptual model to be tested; (2) specific hypotheses to be corroborated or falsified by the experiment; (3) operational definitions of the geomorphological properties and measurements of interest; (4) a formal schedule of measurement, together with appropriate controls; (5) a specified scheme of analysis; and (6) data collection and a management system designed for the task at hand. Pursuit of the two paths advocated by

Nystuen and Church would do much to accelerate the rate of progress in geomorphology.

Geomorphologists have founded their discipline on analogies, so it is appropriate to end Part One of this text with another analogy. Men live by virtue of oxygen, water, and food; geomorphology survives by theory, fieldwork, and a contribution to society at large. Life without oxygen is fleeting, and without theory so is geomorphology; life without water is a brief, lingering passage as would be geomorphology without field verification of theory (remember where uncorroborated theory put Lord Kelvin in the storybook of Earth history!). Life without food precludes growth; geomorphology without something to say to society at large will never grow. This does not mean that applied geomorphology in the usual sense will save geomorphology. At present, it seems that applied geomorphology (as usually conceived) is doing as much to transform the discipline by restricting it to a human, management scale as it is to preserve, or foster it. Geomorphology (i.e., the study of all landforms, together with their initiation, development, and degeneration processes) needs to exhibit better theoretical articulation so that it may initially present a more persuasive case to the scientific community at large, and, thereby, subsequently become more visible to society in general.

Part two

10 Traditional models of landscape evolution

Thus far only individual threads in the fabric of geomorphology have been examined. Some are colorful and have attracted much attention, while others are woven deeply and although rarely seen may be compared to the backing of a carpet in that they hold things together. Regardless of their individual roles these ideas truly are threads because they must be woven together before the entire pattern of geomorphology can emerge. Furthermore, the analogy may be extended because individual threads may be woven in a variety of designs at the weaver's discretion.

The weaving analogy may also be used to characterize Part Two of this text. Attention will now be turned away from examination of individual threads to the designs that have been presented for the entire fabric of geomorphology. If perfect, such a design should permit both retrodiction and prediction of landform development, as well as embrace the entire range of landform scales. Part One should have shown the magnitude and complexity of such an undertaking; nevertheless it is one that has been attempted on a number of occasions and in a variety of ways. In this chapter, three models that pre-date and/or exclude the use of quantitative data are reviewed.

William Morris Davis

It is fitting to begin any review of landscape models with Davis's geographical cycle. While it was not the first attempt to create a universally (or at least widely) applicable model of landscape development, it is probably fair to claim that it was the first such model to gain widespread acceptance within the discipline. Furthermore, its longevity has been quite remarkable. The geographical cycle no longer dominates research thinking as it did in its heyday, but it remains widely used as a teaching tool. In addition, the persistence with which geomorphologists cling to cyclical models suggests that the geographical cycle retains residual influence. The review here comprises a brief description of the scientific mood during the period in which Davis formulated the model, a synopsis of the model, and an evaluation.

Scientific context

Davis's publications span the period 1880–1938, with those papers published later in his life revealing a much greater willingness to reconsider the universal applicability of the geographical cycle than is generally realized. Nevertheless, Davis's publications between 1883 and 1899 embrace the crux of the ideas for which he is remembered; while the volume *Geographical essays* (Davis 1909a) may be viewed as the "bible" from which generations of Davisian geomorphologists were taught.

It is hard to underestimate the impact of Darwin's (1869) *On the origin of species*, first published in 1859, on the scientific community of Davis's most fruitful period. Having himself been strongly influenced by Lyell's (1830–33) *Principles of geology*, Darwin embraced the notion of incremental change over time (Stoddart 1966, pp. 685–6). Stoddart suggested that this is perhaps why time is the dominant theme in Davis's model. Both Chorley (1965, pp. 29–32) and Stoddart (1966, p. 688) pointed out that "evolution" (a term that emerged in the fifth edition of *On the origin of species*) was for Darwin a process, but for Davis and many others merely history. In other words Darwin was preoccupied with the mechanism of change, but was thwarted prior to Mendel's work on genetics, while Davis and others embraced evolution as a notion of inevitable change, development, or progress over time. This shift from Darwin's perspective contained another important change (Stoddart 1966, p. 695), because Darwin initially cast natural selection in a probabilistic mold, yet geographers, including Davis, interpreted it deterministically.

The period of Davis's early writings was also one during which important new ideas on crustal behavior emerged: Suess (1883–1908) published his eustatic theory, Dutton (1889) named isostasy, and Gilbert introduced the concept of epeirogeny. While Davis did not embrace all of the new concepts (e.g., Suess's eustatic theory), his writings clearly reflect his familiarity with both established wisdom and new developments in geomorphology.

The geographical cycle

The concepts inherent in the geographical cycle, Davis's model of landscape development, pervade all of his geomorphological publications subsequent to its initial presentation in 1884. As Davis constantly defended, restated, and modified the model, a full grasp of all its nuances can only be obtained by reading virtually all of his papers, or at least the exhaustive biography by Chorley *et al.* (1973). The presentation here will encompass all of the salient attributes presented in the definitive paper entitled "The geographical cycle" (Davis 1899), plus consideration of the major modifications that followed. Those who wish to go beyond this can do no better than start with the biography cited above, while King & Schumm (1980) provided a much more personal, but extremely interesting, account of Davis's ideas.

Davis wanted to present a model of landscape development that was deductive and theoretical (Davis 1899, pp. 483–4). In public, Davis claimed that his model was purely deductive, that is, it derived from his initial assumptions and not from his field experience. However, in his private correspondence (cited by Chorley *et al*. 1973, p. 549) Davis conceded that his claim of deductive purity really reflected his personal preference for deductive presentations, rather than a truly exclusive use of this approach in his personal research. Certainly later researchers have not been convinced of Davis's deductive purity (e.g., Dury 1969, p. 3). Despite this limitation, and as Chorley (1965, p. 22) commented, it is a real measure of Davis's intellect that after he created the model he was able to state that he could envisage many more landforms than those for which he could find field examples. In addition to a theoretical model Davis (1899, p. 484) wanted one that was genetic, that is, he wanted not only to describe landforms but also to explain them. These theoretical–genetic aspirations were stated very fervently in Davis's writings and serve to remind us that he was not only a great geomorphologist but also a man who consciously tried to make a methodological contribution (i.e., influence the manner in which research is undertaken) to geography at large.

The objectives outlined above were pursued through analysis of the impact of variable structure, process, and time upon landscape morphology. As has already been suggested, time was regarded by Davis (1899, p. 482) as the pre-eminent factor among the three. In truth, this list of basic variables is a truncated or simplified one fostered by Davis himself. In his initial presentation of the model, Davis (1885) identified the rate of uplift as an important variable (Chorley *et al*. 1973, pp. 164–5); furthermore, as Beckinsale (1976, pp. 449–50) has pointed out, Davis (1909b) claimed that his model actually embraced five factors (structure, process, stage, relief, and texture [of dissection]).

Structure for Davis was centered upon the notion of regional structure, such as a plateau region of near-horizontal strata or a mountainous zone of disturbed crystalline rocks. The origins of such zones lay beyond his immediate interests and he took such attributes as initial inputs to his model. Process was an umbrella term under which he recognized the existence of most weathering and transport mechanisms with which we deal today. However, Davis (1899, p. 482) took "weather changes and running water . . . as . . . a normal group of destructive processes". Time as a concept was expressed not in years but in a purely relative sense in which the extent of landscape development was expressed relative to that which could be anticipated upon completion of an entire "geographical cycle" (Davis 1899,

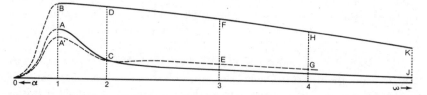

Figure 10.1   W. M. Davis's original schematic outline of surface lowering during the geographical cycle. (From Davis (1899).)

p. 483). This is the concept of "stage(s)" of which the most famous are youth, maturity, and old age.

The kernel of Davis's model appears in Figure 10.1. Time elapses along the horizontal axis ($\alpha\omega$), and the vertical axis represents elevation above sea level. Upon uplift the higher parts of the landscape average elevation B, the lower parts average elevation A; therefore, AB represents average local relative relief. Initial (consequent) streams have elevation A, but by period 2 have reduced the elevation of their channels to C. During the same period higher interfluves are reduced more slowly (only to D) and relative relief is increased (i.e., CD > AB).

Subsequently, main streams lower their channels slowly (CEGJ), but uplands are reduced more substantially (DFHK) due in large part to dissection by headwaters. Upland reduction attains a maximum in periods 3 and 4, unlike valley deepening which is at a maximum in periods 1 and 2. While relative relief experiences a maximum rate of increase during periods 1 and 2 it attains an absolute maximum during periods 2 and 3, when landforms also display their greatest diversity. During periods 3 and 4 relative relief decreases at its fastest rate (but this rate of change is a much slower one than that of increase during periods 1 and 2) and slopes become gentler. From period 4 onward change is extremely slow and timespans very great; in addition slope angles are very low and the landscape is reduced to a rolling lowland regardless of its initial state.

While Davis was well aware that his model was a simplification of reality, he amplified it in many ways, not least of which was by the creation of a wide array of terms, some 150 or more (Beckinsale 1976, p. 455). In presenting an expanded version of the cycle it is practicable to use some secondary sources, and what follows is a synthesis drawn from standard presentations and discussions of the cycle.

Davis pruned his model vigorously (Chorley *et al.* 1973, p. 195) in order to highlight the changing geometry of erosional landforms at different stages (his purely relative and truncated version of time). Despite pruning, the model remained comprehensive because, as Harris & Twidale (1968, p. 238) pointed out, it really represents three models: a cycle of landscape development; a cycle of river development (both individually and as networks):

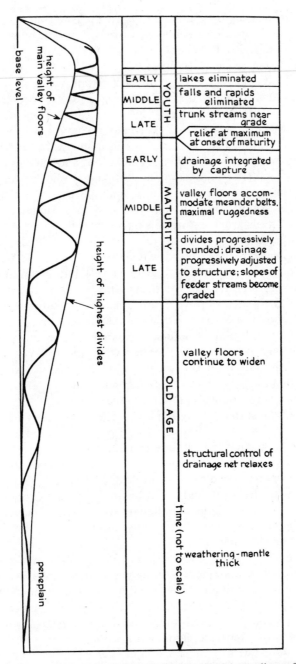

Figure 10.2 An elaborated schematic of W. M. Davis's outline of landscape lowering during the geographical cycle (adapted from work of W. M. Davis, D. W. Johnson, and O. D. von Engeln by G. H. Dury). (From Dury (1969).)

and a cycle of slope development. Each cycle is centered on the stage concept, which in turn is defined by various form attributes. A very brief summary of each developmental cycle is in order here.

The landscape cycle (Fig. 10.2) begins with a brief period of youth, beginning at the cessation of uplift and ending when all of the original uplifted surface is removed (i.e., when valley walls on opposite sides of interfluves have eroded so greatly that ridge crests are below the original, uplifted surface). This formally defined boundary between youth and maturity is not matched by one between maturity and old age, which is simply a transitional phase. Old age is very long and can only be terminated by renewed uplift, heralding the onset of a new cycle and reappearance of youth. A landscape may experience multiple cycles and, therefore, its surface may exhibit youth, maturity, and old age simultaneously, but at different locations; such a surface is normally called polycyclical.

A landscape exhibits increasing relative relief during youth (see above) so that it is at a maximum at the onset of maturity. Valleys are V-shaped, deep if the region is high above sea level (base level), but shallow if the elevation is close to sea level. During youth, floodplains are essentially absent, but interstream areas are extensive. During maturity, relief attains its maximum ruggedness because of extensive fluvial dissection; this is characterized by sharp divides and minimal interstream areas. As maturity advances the divides become progressively rounded while the valley system becomes more extensive, adjusted to structure, and individual valleys widen so that there are floodplains with moderate meander belts.

During old age, relative relief is greatly reduced, as is the overall elevation of the surface above sea level. The surface is gently rolling, with very broad valleys incorporating wide meander belts and low longitudinal gradients. These trends, if uninterrupted, culminate in a "plain without relief" (Davis 1899, p. 497). However, Davis chose to publicize his penultimate landform: ". . . an almost featureless plain (a peneplain) showing little sympathy with structure, and controlled only by a close approach to base level, must characterize the penultimate stage of the uninterrupted cycle" (Davis 1899, p. 497). Such a surface might still exhibit upstanding erosional remnants known as monadnocks.

Davis's model of landscape development was fully matched by a description of drainage development. During youth, there are a few major rivers, as well as numerous minor, but few major, tributaries. Floodplain development is rare, but there are extensive areas of poorly organized drainage. Maturity is marked by a well-integrated drainage system with widespread development of moderate floodplains within valleys. During old age, valleys are broad with gentle slopes and very extensive floodplains. Meandering is extensive and the overall power of fluvial processes becomes secondary to that of mass wasting.

Davis wrote relatively little that was devoted exclusively to slope devel-

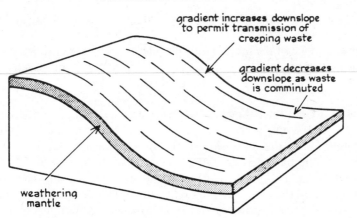

gradient increases downslope
to permit transmission of
creeping waste

gradient decreases
downslope as waste
is comminuted

weathering
mantle

Figure 10.3 The classic sigmoidal, convexo–concave graded slope produced during the Davisian geographical cycle. (From Dury (1969).)

opment, but contributed considerably to ideas on the topic in his many reiterations of the geographical cycle (see Young (1972, pp. 25–8) for a succinct summary). The single most important concept that Davis contributed to ideas on slope form was that slope angles decline through time. This was founded on two fuzzily entwined ideas: (1) particle size decreased downslope and consequently so does the angle necessary to maintain movement (an argument parallel to the stream one); (2) more material is removed high on a slope than at the toe. The first argument centers on weathering rate, the second on rate of removal (Young 1972, p. 27).

Davis (1892) proposed an explanation of the upper convexities found on badland slopes. His argument was founded on the notion that creep processes dominated in this position. Although Davis restricted his idea to explain what he considered to be an exception to Gilbert's (1877) "law of divides", Gilbert (1909) himself accepted Davis's explanation and incorporated it into an influential paper devoted entirely to the topic of upper hillslope convexities (Chorley *et al.* 1973, p. 151). In fact, this idea has withstood the test of time and Carson & Kirkby (1972, pp. 299–300, 437–9) showed not only that it is of much broader applicability than Gilbert realized but that it may be substantiated mathematically.

Davis also forwarded an explanation of the basal concavity found on many slopes using the above-mentioned idea that slope materials become finer downslope (Davis 1932). Synthesis of his concepts concerning upper convexities and lower concavities on slopes produces a sigmoidal, convexo–concave slope profile (Fig. 10.3) with both convex and concave elements expanding through time. Both slope angle decline through time and overall convexo–concave form have been fundamental debating points in slope geomorphology for many years, particularly with respect to whether they are of restricted or universal applicability.

Interwoven into the three developmental cycles are some individual concepts that have themselves attracted much attention and deserve brief recognition. One of the most enduring and hotly debated of these ideas was that of stream grade. Davis (1902, p. 86) defined the term "grade" in the following way:

> . . . we find that the balance between erosion and deposition, attained by mature rivers, introduces one of the most important problems that is encountered in the discussion of the geographical cycle. The development of this balanced condition is brought about by changes in the capacity of a river to do work that the river has to do. The changes continue until the two quantities at first unequal, reach equality; and then the river may be said to be graded, or to have reached the condition of grade.

The idea that there is some sort of balance between a stream's energy and its erosive and transporting power is an extremely old one, which Knox (1975, pp. 171–3) was able to trace back to at least the late 1600s. However, the traditional approach had been to interpret the balance (clearly an equilibrium concept) as being between contemporary processes and stream energy. Therefore, the important change made by Davis when he redefined grade (an existing term at the time) was that it was something that a stream attained only over long periods of time. Using the temporal terms of the geographical cycle, grade was absent during youth, achieved during maturity, and sustained thereafter. Rivers exhibiting grade were assumed to have a smooth, concave, longitudinal profile controlled overall by the ultimate base level (sea level) and locally by temporary base levels (see Davis (1902) for a discussion).

Grade became a vital concept in geomorphology, but one that was rather loosely defined. Eventually, the desire to distinguish between an ungraded and graded state led to the concept coming under fire. Mackin (1948) made a famous attempt to redefine the term in an acceptable fashion but failed, largely because he placed too much emphasis on stream gradient. A little later Leopold & Maddock (1953) attempted to identify quantitative criteria within hydraulic geometry with which to distinguish between an ungraded and graded state. Their failure to do so marked the end of a preoccupation with Davis's concept of grade. Dury (1966) and Knox (1975) provide comprehensive appraisals of the debate concerning grade for those who wish to pursue it more fully.

Davis (1899, p. 495) used the term "grade" in a similar, but separately defined, fashion to describe a hillslope condition:

> A graded waste-sheet may be defined in the very terms applicable to a graded water-stream; it is one in which the ability of the transporting

forces to do work is equal to the work they have to do. This is the condition that obtains on those evenly slanting, waste-covered mountainsides which have been reduced to a slope that engineers call "the angle of repose, . . ."

Similarities between Davis's concepts of stream and slope grade were extremely strong because he believed that slope grade extended from slope toe upward (just as stream grade expanded upstream from the mouth). In addition, both a graded stream and a graded waste-sheet could exhibit lower slope angles through time, as in both instances finer material would be moving through or across them as time elapsed. However, slopes and streams were not expected to attain grade at the same time.

As is well-known Davis rooted his model in a "normal" climate, defined as a mid-latitude temperate one. Presumably, Davis hoped that his norm would be a universally valid one, or, at the least, of widespread applicability. Limitations actually appeared quite quickly; Davis (1905) himself published an "arid cycle" and a steady trickle of other special cases followed, with Peltier's (1950) "periglacial cycle" apparently being the last.

Extensive Davisian treatment of geomorphology appeared in such texts as *Geomorphology* (von Engeln 1942) and *Principles of geomorphology* (Thornbury 1969). The latter is interesting because Thornbury's presentation not only provides a good summary of Davis's ideas, but also clearly reflects the struggles experienced by geomorphologists as they became skeptical of Davisian concepts, but were unable to identify an equally comprehensive, alternative scheme. In fact, criticism of the Davisian scheme had always existed, but its various protagonists were rarely able to muster much success in the English-speaking world. However, by the 1930s and 1940s various attacks began to make some impact (see Chorley *et al.* (1973) for a discussion). Instead of following these developments chronologically, only an integrated synopsis will be offered here.

Davisian geomorphology: strengths and weaknesses

The strength of Davis's ideas cannot be substantiated any more firmly than by noting their pervasiveness and longevity. The fact that they still merit attention, above and beyond that assigned to a historical footnote, 100 years after their embryonic emergence is a truly remarkable measure of Davis's intellect. However, the geographical cycle no longer dominates geomorphological thinking as it once did. In short, it is possible to see both strengths and weaknesses in the model. Comprehensive discussions of the appeal and limitations of Davis's ideas may be found in Chorley (1965), Flemal (1971), Chorley *et al.* (1973), and Higgins (1975).

Chorley (1965, pp. 22–3) suggested that the innate appeal of the geographical cycle is its theoretical tenor, reinforced by its simplicity. Higgins

(1975, pp. 12–18) supplied a longer list of appealing specifics, namely: simplicity, applicability, elegance of presentation, seemingly careful fieldwork, filling of an intellectual void, synthesis of prevailing thought, predictive and retrodictive, rational, embracement of evolution, confirmation of prevailing stratigraphic concepts, innate appeal of humid climate being "normal", and cyclicity. In brief, it was a model that not only answered the questions being asked by geomorphologists at the time it was introduced, but did so by appealing to principles embraced by virtually the entire scientific community of the time.

It is easy to believe that Davis's powerful personality, extensive travel, widespread lecturing, numerous publications, active correspondence, and position at Harvard University all served directly to maximize the impact of his ideas (Chorley et al. 1973). Indirectly, the geographical cycle was also served well by its promotion by several important figures in the international geomorphic establishment, notably by D. W. Johnson in the U.S.A., S. W. Wooldridge in Great Britain, H. Baulig in France, and C. A. Cotton in the Southern Hemisphere.

One other strength of Davisian geomorphology merits mention. Davis was a gifted artist and his field sketches did not merely illustrate his papers, but rather conveyed the entire message in an alternative medium to the written word. They are generously illustrated in King & Schumm (1980). This may seem a trivial point but, if one considers the faith contemporary geomorphologists appear to have in slides, it is easy to envisage the powerful form of communication sketches represented in an earlier time. A sketch also contains the inbuilt quality of being editorial in and of itself.

Hindsight invariably seems better founded than foresight. Consequently, it is not surprising that as the discipline of geomorphology has grown, so has an appreciation of Davis's errors, both real and merely perceived. In reviewing the limitations of Davis's work it is important to see them in perspective and not to become so enamored of them that one loses sight of the enormity and longevity of his contribution.

Chorley (1965, p. 28) considered one of the pervasive errors in Davis's work to be the linkage of various aspects of the landscape with single causes, or at most two or three. According to Chorley (1965, p. 24) Davis tacitly embraced the idea that "... the amount of energy available for transformation of landforms is a simple and direct function of relief or of angle of slope". Such an assumption is very much at odds with present-day concepts in process geomorphology, which tend to emphasize complexity and interaction. In large part, Davis's nonquantitative approach may be seen as responsible for such errors. Indeed, his failure to undertake quantitative measurements is very much at odds with contemporary concepts of what constitutes science.

Preoccupation with stage became the hallmark of Davis's model of landscape development. However, this was entwined with something that was not inherently part of the cycle and that was denudation chronology. In denudation chronology the erosional development of a region is related to changes in base level. Such a methodology was already established when Davis presented his cycle, but his approach was so compatible with the existing aims that the two quickly developed into a virtually inseparable entity (Chorley 1965, pp. 33–4). As a result regional denudation chronologies became the primary vehicle of Davisian geomorphology. In fact, the geographical cycle became so entrenched as the foundation upon which regional denudation chronologies were developed that considerably more research energy was spent on debating the regional histories (particularly the number of cycles experienced) than was ever spent on testing the geographical cycle itself. The work of S. W. Wooldridge in Great Britain may be taken as an extreme example of this trend.

As stage and denudation chronologies occupied the limelight in Davisian geomorphology, so structure and process were overshadowed. Little can be said about this save to point out that Davis himself did little to pursue a balanced approach to the three principal factors he had identified. It is no accident that rejection of Davisian ideas produced an emphasis on process geomorphology; the geomorphological community merely became pre-occupied with what Davis and his followers had left undone.

The combination of rapid uplift, followed by long crustal stand stills, and hence constant base levels, was also attacked. Much of this debate is subsumed in the Davis–Penck conflict and follows in the next section. However, the issue of relatively rapid uplift and relatively slow erosion has never been fully resolved; indeed, Schumm (1963b) used available uplift and erosion rate (i.e., quantitative) data to show that Davis's proposal is probably a reasonably sound approximation. Nevertheless, the peneplain is a concept that has generally not fared well in recent decades. As Davis specified that a poor relationship should exist between regional rock structure and a peneplain, it has usually been identification of some sort of structural influence that has most commonly led to reinterpretation of peneplains in non-Davisian terms. In fact, today there is no widely accepted example of a peneplain cited in the literature. Hack (1960), noting the frequency with which the occurrence of "maturely dissected" landscapes were reported in the literature, suggested that they, rather than peneplains, may be the equilibrium form (Hack's ideas will be examined in the next chapter).

Davis launched his model by considering a "normal" climate. This was defined as a humid, mid-latitude, temperate environment and inevitably focused attention on fluvial geomorphology. Initially, the notion of a normal environment was not questioned directly, although the rapid development of a number of other Davisian or Davisian-like cyclical models for differing climatic regimes clearly suggests that the norm was not as widely applicable as Davis had intended. However, much later, the choice of a humid, temperate regime as an appropriate worldwide norm came under direct fire. This debate came in two quite distinct forms. First, Lester C. King (e.g., King 1957, 1963a) forwarded the argument that Davis simply chose the wrong kind of norm; as King's model of landscape development is to be examined directly later in the chapter, this issue will not be taken up here.

The second objection to Davis's model of Appalachian development comes from a number of semi-independent concepts which share the basic notion that the Appalachians may actually have developed under the influence not only of fluvial processes but also of other important factors. As late as the mid-1960s Judson (1965) noted that Davis's ideas of Appalachian development still dominated thinking about the region. However, other explanations of the region's landscape exist: Denny (1956) emphasized structural control; Hack (1960, 1980, 1982) invoked contemporary processes, especially differential erosion and ongoing tectonics; Judson (1975) has also emphasized ongoing tectonics; glaciation is known to have occurred in some sections of the Appalachians (Marchand 1978); and Ciolkosz (1978) demonstrated that periglacial forms are widespread. Whatever final assessment emerges it is unlikely to be founded on the premise that the Appalachian regional landscape developed under an exclusively fluvial regime.

Remarkably few of Davis's process concepts have withstood the test of time; undoubtedly his failure to make field measurements has much to do with this. Perhaps, the most important of his process concepts to fall by the wayside was grade, because it was central to his model and was for so long a central issue of debate. Nevertheless, most of Davis's other concepts concerning fluvial processes have also faded, generally to be replaced by explanations founded on hydraulic theory.

Davisian geomorphology now stands in a peculiar position. In part it has been proven wrong, but in part it has been bypassed rather than replaced. Davis's failure to undertake process measurements undoubtedly stimulated the modern emphasis on quantitative, process geomorphology. However, while the initial impetus to pursue process geomorphology was to seek redress of the then prevailing imbalance, this very step changed the scale at which geomorphology was pursued. Process studies cannot be undertaken at a regional scale (except through elaborate statistical structures) because of obvious fiscal and logistical constraints. Therefore, process geomorphologists have ended up answering different questions at different scales from those posed by Davisian geomorphologists. During the upsurge of process geomorphology remarkably few geomorphologists have continued to pursue their trade at the regional scale. Consequently, many of Davis's ideas simply hang in suspended animation; they are largely in disfavor, but have never actually been addressed comprehensively, let alone disproven. Even at its zenith the geographical cycle did not go unchallenged and it is to two of the most serious challengers that attention will now be directed.

Walter Penck

Walter Penck's contribution to geomorphology is truly enigmatic. This stems from three basic difficulties: (1) his primary work, *Die morphologische Analyse* (Penck 1924), was published after his death; (2) his writing style is extremely difficult to understand; and (3) he was seriously misrepresented by W. M. Davis, whose 1932 translation/interpretation of Penck's ideas was the primary source in English until 1953.

Given Penck's complex writing style it is not really feasible for the average academic equipped with basic German language skills to read or translate the original text and grasp the full range of subtleties involved. Therefore, it seems better to rely on Czech & Boswell's (1972) translation (the first edition of which was published in 1953) than to run the risks associated with an amateur translation. Nevertheless, it is important to appreciate that this is a significant limitation. References to the Czech & Boswell translation will follow the convention used by Simons (1962) in his article on Penck (i.e., M.A. 100 refers to p. 100 in Czech & Boswell's (1972) English version of *Die morphologische Analyse*).

Scientific context

There is no readily available source in English from which to derive the general scientific influences on Walter Penck and he has not been the focus of much attention in his native Germany. What follows has been gleaned from the limited English-language sources that deal directly with Penck himself. Walter was the son of Albrecht Penck, one of the most famous geomorphologists of his time (Chorley *et al*. 1973, pp. 515–36). Albrecht Penck wrote a major text on general geomorphology and was an authority on glaciation in the Alps. His influence was also greatly enhanced by the importance of the academic positions he held, as well as by the large number of his talented students who became well-known geomorphologists in Germany (Bremer 1983). Albrecht Penck and William Morris Davis shared a long professional and warm personal relationship, which then changed into a fierce intellectual clash and personal disaffection. The turning point was Davis's (1912) publication in German and the disaffection was reinforced by the two men being on opposite sides in World War I (Chorley *et al*. (1973) provide an interesting selection of the correspondence between the two).

It appears that Walter enjoyed and suffered many of the classical outcomes of being the son of a famous father. He was very well traveled at an early age and obtained his doctorate at 22. He must have known Davis quite well as a child, as the Penck and Davis families visited each other and corresponded fairly frequently. For two years Walter served as a geologist in Argentina, and subsequently as an academic in Turkey. Walter enjoyed a close relationship with his father and it seems likely that given the eminence of the latter he must have been a profound influence. However, this cannot have been overwhelming or stifling because Bremer's (1983) comparison of their two methodologies reveals strong contrasts. Following World War I, Walter was only able to obtain minor academic appointments and, there-fore, completely lacked the power and influence of his father. In 1923, before publication of his major works and when he was aged only 35, Walter died.

Penckian geomorphology

Walter Penck was interested in tectonics or crustal movements and attempted to approach this topic by building a theoretical model relating surficial landforms to subsurface movement for the purpose of explaining the latter (M.A. 6). This sphere of research he dubbed "morphological analysis" stating: "Morphological analysis is this procedure of deducing the course and development of crustal movements from the exogenetic pro-cesses and the morphological features" (M.A. 6). In a letter to Davis (Chorley et al. 1973, p. 541) he wrote "... the shape [Gestalt] of landforms [Landformen] is a function of the ratio of the velocity of endogenous movement to intensity of erosion". Characterizing the issue even more tightly, Penck (M.A. 6) pointed out that endogenetic processes, exogenetic processes, and landforms may be viewed as an equation in which the last two may be examined, and therefore are knowns, and may be used to derive the third and unknown factor – endogenetic processes. Geomorphology was secondary for Penck in two ways: (1) he believed that endogenetic processes controlled exogenetic processes (M.A. 11); (2) he turned to geomorphic modeling to supplement his tectonic investigations only after his primary tool, stratigraphy, proved inappropriate in some locations (Chorley et al. 1973, p. 698).

It is essential to appreciate Penck's objective because it obviously dictated many of the intellectual structures he chose to create. There are some limited similarities between Davis's and Penck's methodologies (Young 1972, p. 28): both made visual observations of slopes and explained them by deductive considerations, but then failed to conduct detailed comparisons between their postulated forms and field data. However, these are the only similarities and the differences loom much larger.

The Davisian model is pervaded by the notion of stage (i.e., how much of the supposed complete cycle has the landscape experienced); the Penckian model is similarly infused by the concept of uplift, most particularly by the

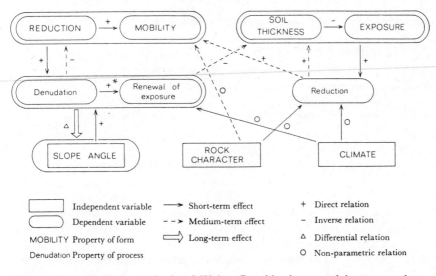

Figure 10.4 Young's synthesis of Walter Penck's slope model, presented as a canonical structure (see Ch. 12). (From Young (1972).)

notion of rate of uplift. It is apparent that, at least initially, Penck saw his model as an extension of Davis's rather than an alternative (M.A. 12, Chorley *et al.* 1973, p. 543). However, the tenor of Penck's writing, both in his text and personal letters to Davis, suggests that this is something of a contrived posture. He stated that the Davisian cycle is valid, *if uplift is followed by erosion on a stationary crust*, but then noted that this is a rare or nonexistent occurrence. Penck then highlighted the fact that Davis himself claimed that his "uplift, followed by erosion" presentation of the geographical cycle was a simplified, special case used for ease of illustration. Why, then, Penck wanted to know, did Davis and his followers discuss landscape development only in terms of this self-admitted, contrived special case? It seems quite likely that Penck sought to minimize the conflict between his own ideas and those of Davis in order to insure that his own were not stillborn, although other explanations for this early posture are possible.

Penck did not see landforms as following predictable sequences through time. Rather he saw them following any number of possible sequences controlled by the changing interplay between uplift and erosional intensity (or rates). These relationships were considered with respect to individual slope profiles, regional assemblages, and river longitudinal profiles. In this sense it is possible to compare Penck with Davis, but his slope concepts were much more highly developed than Davis's and his ideas about mass wasting very advanced for the times (Young 1972, p. 29).

Clearly, Penck had to create a comprehensive model of exogenetic processes in order to gain his objectives. The manner in which he did this, although flawed, was sufficiently rigorous that Young (1972, pp. 29–34) was able to summarize it in a form that is very close to a modern process–response model (Fig. 10.4). Penck (M.A. 16–18) carefully specified the domain of his model, excluding submarine environments, eolian processes, and glacial landscapes. Furthermore, he eliminated climatic variability as a source of landform variability (M.A. 17 and 72) and postulated (M.A. 120) that similar forms appear in different climates, but at different rates. Bearing in mind these concepts it is useful to follow Young's summary of Penck's exogenetic model, which is paraphrased below.

Seven form properties were recognized by Penck as follows:

(1) *Degree of reduction* – The extent to which the regolith (soil in Czech & Boswell's translation) is comminuted. The reduction depends directly upon the rate of weathering and inversely on the rate of denudation (see special definition below).

(2) *Mobility* – The ease with which denudation may remove regolith. There is a critical value of regolith size for each slope angle, and this must be attained to mobilize the regolith. There is a direct relation between reduction and mobility, and a nonparametric relationship between rock type and mobility.

(3) *Regolith thickness* – Thickness depends directly on the rate of reduction and inversely on the rate of denudation.

(4) *Bedrock exposure* – The degree to which bedrock beneath the regolith is exposed to the reduction process.

(5) *Rock character* – This nonparametric concept embraces all rock properties that influence the rate of reduction and regolith mobility. It is an independent variable.

(6) *Climate* – An independent, nonparametric variable that is assumed to influence rates of reduction and denudation.

(7) *Slope angle* – This begins as an independent variable, but becomes dependent upon the reduction rate. It has a differential relationship, a downslope increase in the rate of denudation increases the slope angle.

The seven form properties were complemented by three processes:

(1) *Reduction* – Breakdown of the regolith into finer fractions by all weathering processes. Reduction is dependent upon rock character and climate in a nonparametric fashion, and also directly upon exposure.

(2) *Denudation* – As used by Penck this term refers to the removal of regolith from a slope surface. Its rate depends on climate nonparametrically, and directly upon both mobility and slope angle.

(3) *Renewal of exposure* – As Young (1972, p. 30) pointed out this is not

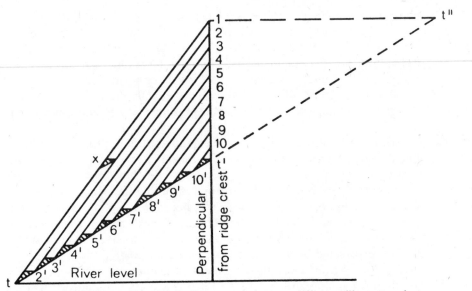

Figure 10.5 Penck's simplified diagram, using a cliff, to illustrate slope replacement. (From Czech & Boswell (1972), after Penck (1924).)

really a process, merely treated as one. It is the exposure of the bedrock surface beneath regolith to reduction. Rate of renewal is directly dependent on the rate of denudation.

These seven properties include three closely related pairs, degree of reduction and mobility, regolith thickness and exposure, plus denudation and renewal of exposure. While Penck understood the independent and dependent nature of the relationships he postulated, Young (1972, p. 30) suggested that he probably did not appreciate that such relationships could change over different timescales. Furthermore, if the process–response model outlined in Figure 10.4 were to be operationalized, rock character and climate would have to be quantified. Finally, Penck made one error concerning the relationship between denudation rate and renewal of exposure (which amounts to slope retreat). The error was that Penck assumed that as soon as regolith reached the necessary size to permit mobility at a particular slope angle the material would be removed. This means that he failed to incorporate the fact that the mobilized material could only move by crossing the downslope surface, thereby causing a decreasing rate of exposure downslope. This represents an external error (i.e., it does not match reality); however, it is not an internal error (i.e., the internal logic of the model remains viable).

Like Davis, Penck then illustrated his ideas with a simplified case, that of a cliff (Fig. 10.5). The cliff has a level surface above and a river at its foot, which removes all the debris it receives, but does not erode the cliff base. The cliff face weathers uniformly and the weathered materials drop off the cliff in a similar fashion. As a result the cliff face retreats parallel to itself, except for the tiny fragment $(t-2^1)$ at the foot where material is not transported because the angle is not steep enough. This sequence is repeated time after time, resulting in parallel cliff retreat and a series of ever-shallower slope segments expanding upslope – given uniform material and crustal stability.

The analysis outlined above is then extended to slopes without cliffs. As noted above, Penck's assumption of uniform, instantaneous removal of reduced debris is suitable for the case of a cliff, but erroneous in other contexts. Furthermore, he failed to appreciate that the period of exposure will decrease upslope from the slope toe, becoming zero at the slope crest. However, these errors do not entirely negate Penck's model, which has two fundamental attributes derived from the ideas presented so far. First, slopes that are unaffected by ongoing crustal movement flatten through time by the creation of ever-lower-angle segments, each developing first at the toe and then expanding upslope (M.A. 138–9). Second, any individual slope segment retreats parallel to itself (M.A. 138–9). Therefore, Penck's model of slope development is best characterized as one of "slope replacement" (Young 1972, p. 34).

Many textbooks still identify Penck as a proponent of the "parallel retreat of slopes"; this is completely wrong and is an error introduced by Davis (1932) that became entrenched in the English-language version of Penck's · ideas. The first real correction that received widespread attention was by Simons (1962), although Tuan (1958) published a short note that clearly demonstrates that the traditional interpretation was erroneous. Thus, any publication in English prior to Simons's paper is likely to have been based upon Davis's (1932) paper, and as a result will be very misleading.

The erosional model summarized above was linked to crustal mobility by several other concepts. Penck believed that, given uniform rock, the rate of denudation is proportional to slope angle (M.A. 71); therefore, steeper slope segments exhibit higher denudation rates and will accordingly retreat upslope at faster rates than slope segments of lower angle. According to Penck, each segment would be initiated at the slope foot and thereafter maintain its angle as it retreated upslope. In addition, Penck thought that in homogeneous materials the steepness of each slope segment was proportional to the intensity of river erosion (M.A. 147). He also considered the rate of upslope retreat of each slope segment to be controlled by the steepest slope angle occurring at the downslope margin of the segment. In Penck's scheme these downslope margins are convex and concave breaks in slope (M.A. 151). Each one of them serves as a local base level for slope segments immediately upslope and thereby isolates these segments from the influence of the main base level, which is the river at the foot of the slope.

Obviously the model is now becoming extremely complex, but it is also providing great flexibility. Its other salient attribute is that each slope provides a record of the variations in the intensity of river erosion, if the intervening complexities outlined above are interpreted correctly. In general terms, concave slope profiles represent a sequence of decreasing intensity of river erosion, while convex slope profiles reflect increasing intensity of river erosion. Penck expanded on this general situation to show that concave profiles would be accompanied by a decrease in relative relief (his waning development; M.A. 153) and that convex profiles would be associated with increasing relative relief (his waxing development; M.A. 155).

The final link in the sequence is to determine the relationship between the intensity of river erosion and the rate of crustal movement. This step was badly short-circuited by Davis (1932) in his interpretation of Penck when he stated that convex slopes form during periods of accelerating land upheaval and concave slopes during periods of decelerating upheaval. In fact, Penck had a typically elaborate analysis of the relationship between river erosion and uplift that was outlined in his text and modified in his later paper (Penck 1925) on the Black Forest. Simons (1962, pp. 9–12) provided a careful appraisal of Penck's statements and it is quite apparent that there are direct contradictions. Penck certainly did not believe that the rate of river erosion is exclusively dependent on uplift rate as he clearly recognized the influence of river discharge and rock type in determining river gradient. Furthermore, he also presented detailed explanations of the formation, upstream migration, and impact on longitudinal river profiles of knickpoints. Despite the many uncertainties that emerge at this point, Penck's model does seem to embrace a general notion of equilibrium between river erosion and crustal mobility – although the inherent conflicts in the argument preclude a satisfactory determination of precisely how this is supposed to operate.

The final component in Penck's presentation is one of generalization. Like Davis he recognized the presence of landform assemblages that constituted typical landscapes, but instead of assigning this similarity to stage, as Davis did, Penck asserted in the preface of his book (M.A. vi) that there is a one-to-one relationship between regional tectonic history and regional landform assemblage. Accordingly, the concept of a norm, such as Davis employed in his model, would stem from tectonic history in Penck's model. In fact, Penck recognized three general categories of landform assemblage: these were summarized by Simons (1962) and it is this analysis that is followed here.

One category is dependent upon folding produced by lateral forces. This class was translated by Davis as a "broad-fold", although Simons (1962, p. 5) suggested that "great-fold" is a more appropriate translation. In modern terminology the closest term is orogenic folding. Penck also claimed (M.A. 28–9) that such folding is widely accompanied by arching of the fold, the latter being dependent not on the folding itself but on intrusion of magma. As an example Penck cited the Alps, with their folding resulting from lateral compression, but their elevation from arching. However, Penck's type site was the Puna de Atacama region in northwest Argentina where he had conducted fieldwork soon after gaining his doctorate. In traditional terms this region is of the basin-and-range type and its origin would be assigned to block faulting; Penck's interpretation was that all such regions are actually the product of great-folding.

Penck's second type of uplift was dome formation without associated folding; in modern terms this is essentially epeirogenic (vertical) movement. His paper explaining the origins of the Black Forest region in these terms is second only to his text (Penck 1924) in importance. In it Penck pointed out that the Black Forest region is a dome bearing a series of roughly concentric erosional benches forming a giant staircase. Such an assemblage was termed "*Piedmontreppen*" and each individual bench was assumed to have been created at the margin of the dome. In this context the center of the dome would be rising rapidly while the margins would rise very slowly at first. This surface at the edge of the zone of active uplift rises slowly, but is nevertheless eroded during uplift and was christened by Penck a "*Primarrumpf*". It is in this context that Penck developed his ideas concerning creation of step-like forms (benches on the land, knickpoints within streams) from continuous uplift, rather than from interrupted uplift as was assumed by those following the Davisian scheme. It is within the argument that Penck used to establish stepped landforms from continuous uplift that many of the conflicting statements mentioned earlier (Simons 1962, pp. 9–12) emerged.

Finally, Penck recognized that extensive regions appear to have experienced crustal stability. Believing that mass wasting alone could only produce concave slope profiles, Penck envisaged such landscapes to be dominated by gentle concave slopes with occasional concave-sided erosional residuals called inselbergs. This issue is also the center of some controversy concerning the nature of slope profiles, but will not be pursued here.

The last attribute of Penck's model considered here is illustration of his incorporation of uplift rates into landscape character. This is succinctly stated in a letter from Penck to Davis (Chorley *et al.* 1973, pp. 544–5). Penck believed that uplift begins and ends slowly (Simons 1962, p. 5). Therefore, he disliked Davis's notion of each cycle beginning with an uplifted peneplain (essentially unchanged from its cycle-ending form). He overcame this difficulty by identifying "*Endrumpf*" (a true terminal peneplain dominated by concave slopes) and "*Primarrumpf*" (a slowly uplifted and erosionally modified "peneplain" dominated by convex slopes). He went on to associate *Primarrumpf* with slow uplift, intermediate forms (Davis's mature forms) with intermediate rates of uplift, and alpine forms

(Davis's young forms) with very rapid uplift. Given his ideas of the temporal variation of uplift rates this produced a spatial pattern with mountainous regions exhibiting a *Primarrumpf* in the central zone, surrounded by intermediate forms, which themselves are surrounded by an outer ring of alpine forms. In essence this model neatly recasts Davis's time-dependent model into Penck's largely time-independent framework.

Penckian geomorphology: strengths and weaknesses

Evaluation of Penck's model is constrained by its hurried writing, posthumous publication, and confused representation in English. As Bremer (1983, p. 135) commented, Walter Penck is essentially an English-language phenomenon, receiving little or no attention in present-day Germany. Whatever its flaws, his model was certainly a massive intellectual achievement.

One of the most important flaws in the model is an external one. Penck based his ideas on the nature of crustal uplift from stratigraphic analysis of the deposits in basins adjoining uplifted blocks in the Andes (Chorley *et al.* 1973, pp. 698–9). At the time this was an accepted approach; today it is considered entirely wrong. As noted previously, his geomorphic model was largely intended as an additional tool to extend his investigation of crustal mobility where stratigraphic evidence was unavailable.

Many of his other ideas cannot be pinned down with precision due either to his complicated writing style (which presumably could be overcome with an excellent knowledge of German and German geomorphic thought of the time), or to direct conflicts in his statements (which presumably are now insurmountable obstacles). However, a number of his ideas remain intuitively appealing (e.g., the relationship between slope angle and "denudation" rate); nevertheless they remain untested quantitatively.

Much of the contempt heaped on Penck's ideas after his death is clearly the result of Davis's poor translation of some of Penck's ideas. In some instances this was exacerbated by Davis's vigorous defense of his own scheme against his own erroneous representation of Penck's! This is why it is so important to appraise Penck from either the original text in German, or from Czech & Boswell's translation and subsequent work in English, rather than from early English-language evaluations. Although, as a matter of principle, any translation not undertaken personally should be treated with caution.

Despite any enthusiasm for the need to treat Penck fairly, it is still probably true that misrepresentation of his work served geomorphology about as much as did his true model. His ideas provided a rallying point for those who disliked Davisian concepts, and nature of uplift was added to structure, process, and time by all but the most entrenched Davisian disciples. Penck's supposed notion of the parallel retreat of slopes focused great attention on slope evolution in a manner that had never been stimulated by Davisian ideas. It is true that one or two notable early researchers (e.g., Grove Karl Gilbert) seem to have resisted the Davisian model completely in their approach to geomorphology. However, prior to Penck there was in the English-speaking world no coherent and comprehensive alternative to Davis's cycle for the vast majority of those who were either being trained as geomorphologists, or were already trained but lacked the rare intellectual talents to create concepts on such a grandiose scale.

Lester Charles King

In recent decades Lester King has stood alone as the geomorphologist of international importance pursuing geomorphology at the global scale reflected in the works of Davis and Penck, although there is recent evidence of renewed interest in this scale of geomorphological studies (e.g., Gardner & Scoging 1983). King's model embraces some components of both of the earlier schemes, rejection of some portions, and incorporation of some altogether new ideas.

Scientific context

King was thoroughly trained in Davisian geomorphology by C. A. Cotton, the leading Davisian disciple in the Southern Hemisphere. As noted by King himself (Higgins 1975, pp. 9–10) he initially tried to interpret the landscape of southern Africa (his primary field area) in Davisian terms and, while this endeavor became less and less satisfactory to him, there is little doubt that his ideas retain some residual Davisian components (King 1963a, p. 723). As the Appalachians influenced Davis, and the Andes and Black Forest influenced Penck, so too has southern Africa made its mark on King's ideas.

King was also influenced by one or two notable South African researchers, most obviously by T. J. D. Fair with respect to slopes and A. L. du Toit on the topic of tectonics. His ideas on slope evolution were also partially derived from those of Kirk Bryan and Alan Wood. Finally it is important to mention the influence of Penck on King, but it must be emphasized that this is a complicated issue. Illustrations of this point include King's acceptance of the erroneous Davisian version of Penck's slope model (King 1963a, pp. 723, 725), but his rejection of Penck's notions concerning relationships between uplift and slope morphology (King 1957, p. 82).

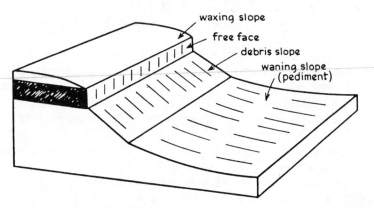

Figure 10.6 An early version of L. C. King's slope model illustrating the four principal components. (From Dury (1969).)

The standard epigene cycle of erosion

Like Davis, King has used a variety of terms to describe his model; in the above title (King 1967) "epigene" simply means "at the surface" or "sub-aerial" (King 1963a, p. 748). King has published extensively over a period spanning some 30 years or more; however, the fundamentals of his model are fully represented in two of his papers (King 1957, 1963a), plus three books (King 1963b, 1967, 1983).

King's model is remarkably comprehensive as it encompasses explanations of the nature of uplift and the cycle of surficial erosion, plus identification of the envisaged surfaces in many parts of the world. Here attention will be focused on the cycle of erosion, and briefly upon the uplift mechanism proposed by King; this means that King's worldwide extrapolation of his model is ignored.

Clearly, the starting point of King's thinking was the large escarpments of southern Africa. Their size, apparent age, and unsubdued forms must comprise a tremendous challenge to even the most agile and gymnastic Davisian mind! King's model is, then, founded on the notion of the parallel retreat of scarps. As mentioned, King incorrectly cited Penck as the source for this concept, but also correctly identified Kirk Bryan as a proponent of the idea. Notice that one critical outcome from an acceptance of the parallel retreat of slopes is that the surface above a retreating scarp does not experience downwearing.

The precise mechanism of scarp retreat proposed by King (see King (1957) for an exhaustive discussion) owes much to slope studies undertaken by Wood (1942) and Fair (1947, 1948a, b). The four slope elements shown in Figure 10.6 and used by King were suggested by Wood (1942). Each element is considered to be semi-independent and, although there is interaction between them, any one of the elements may be entirely absent on a given slope (King 1957, p. 83). Each one will be discussed separately here.

The waxing slope is a convex segment at the crest of the slope. It is considered to be mantled by weathered material and transport across it to be dominated by soil creep. The convexity is believed to exist because material for transport is being generated everywhere on the slope. As a result each point on a downslope profile must transport slightly more material than the adjoining point upslope. It is argued that this is achieved by an increase in slope angle (King 1963a, p. 729); a sequence of increasing slope angles is obviously the essence of convexity.

The free face (synonymous with "cliff") is a bedrock outcrop. It is assumed to retreat parallel to itself under the influence of weathering processes and uniform removal of weathered debris. This limited concept of parallel retreat, restricted to free faces, is correctly assigned to Penck. King (1963a, p. 728) conceded that cliff-foot burial by an encroaching talus is possible, but claimed that debris slopes commonly occupy a constant portion of the slope, that is, they retreat with the scarp rather than burying it (King 1963a, p. 728). The free face is considered dependent upon adequate relative relief, which means that in areas of little local relief it may be absent.

The debris slope is entirely dependent on the presence of a free face for its existence and consequently must be absent when the free face is. According to King (1963a, p. 728) the angle of debris slopes is controlled by fragment size. Therefore, as the debris standing at the angle of repose undergoes further breakdown it is possible for the angle of debris slopes to decrease toward their lower limits (King 1963a, p. 728).

The final element of King's standard slope sequence is the waning slope. It exhibits a gentle, concave profile that is assumed to be a hydraulic one. Below such a surface King (1963a, p. 729) noted that there may be a vertical profile of soil, regolith, then bedrock, or a profile comprised of transported debris resting directly upon an eroded bedrock surface. The latter combination is the classical one for a pediment; it is also possible, although it cannot be proved, that the former combination merely represents a pediment surface that has weathered. The precise processes of pedimentation have long been debated without any real resolution, although some combination of weathering, rill wash, sheetwash, and associated processes is commonly cited. However, King (1963a, pp. 732–5) made a specific version of pedimentation vital to his model of landscape development.

The upper limit of pediments, almost without exception, makes a very sharp angle with the slope element above it. The origin of this so-called piedmont angle has also been the subject of many suggestions. King's explanation is that water traveling across the steeper, upper slope elements in a storm exhibits the usual turbulent (and therefore erosive) flow found in streams. However, when the water reaches the upper limit of the pediment (with a much lower slope angle) he believes it flows in thin sheets exhibiting laminar (nonerosive) flow. Lower down on the pediment, the continued accumulation of rainfall increases discharge to the point where laminar flow once again breaks down and turbulent flow resumes. This combination of flow characteristics, if valid, would account for both the piedmont angle and the gentle concave (hydraulic) curve found across most pediments. Occasional disruptions of this pattern on pediments (e.g., erosion of hollows or basins) are attributed to local factors overwhelming laminar flow and erosive, turbulent flow ensuing.

Several critical concepts emerge from this outline of slope behavior. Parallel slope retreat results in emergence and expansion of the pediment. It also implies that pediment surfaces experience only minimal erosion or reworking once formed. The sum result of these ideas is that in the advanced stages of erosion the landscape will be composed of extensive concave surfaces (coalesced pediments) with isolated erosional remnants. These erosional remnants will have been reduced by erosion of their margins (parallel slope retreat), but the surface above the margins will have experienced little or no erosion because it is part of an earlier pediment surface.

The process of pediments expanding and uniting is called pediplanation, and the resulting landform is a pediplain. It follows from the description of formative process that all pediments and pediplains are time-transgressive (i.e., their surface is not all of one age). It also follows that pediplain surfaces are dominated by concave slope profiles and that the erosional remnants (the conceptual equivalent of Davis's monadnocks) found upon them are steep-sided (King 1963a, p. 750). King's erosional remnants are often called inselbergs; however, the term "inselberg", "bornhardt", and "koppie" have been used in overlapping and somewhat conflicting fashions by a variety of workers. It is possible for the conceptual equivalent of Davis's monadnock to be an inselberg, but many inselbergs and bornhardts are the expression of structural differences during normal erosion and not indicative of the end of an erosion cycle (King 1963a, p. 749).

King (1963a, p. 747) retained the Davisian trio of process, stage, and structure, specifically assigning them the order of importance listed here. However, King rejected climatic geomorphology and believes that all landscapes are essentially similar. As an illustration of this he categorized inselbergs and monadnocks as homologous (i.e., fulfilling the same role). Having offered a concept of unified landscape development within all climatic zones, King made another far-reaching modification by suggesting that it is the semi-arid, and not the humid temperate, regime that is normal. He based this claim of normality both on spatial grounds (the proportion of the Earth's surface currently experiencing such a climate) and on the basis of what has been the most prevalent climatic type through geological time. One important outcome of this shift is that the classical convexo–concave

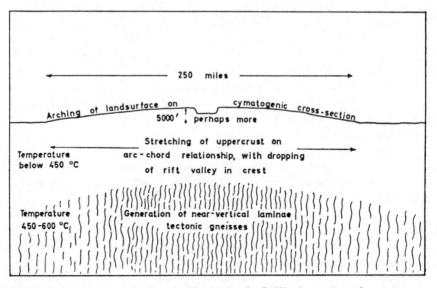

Figure 10.7   A schematic diagram illustrating L. C. King's concept of cymatogeny. (From King (1983).)

slope profile sequence of Davis, which appears in many mid-latitude environments, is defined as degenerate, or largely inactive, as it lacks the dynamic free-face slope element (King 1963a, p. 748).

At larger geomorphic scales King (1963a, p. 749) suggested that his pediplain surfaces may advance across the landscape either by parallel retreat of scarps or by headward advance of nickpoints through drainage systems. The existence of pediplains with their dominantly concave slope profiles means, for King (1963a, p. 749), that Davis's peneplain dominated by convex slope forms is a nonexistent landscape. According to King where such convexities exist they serve as verification of Penck's distinction between an *Endrumpf* and a *Primarrumpf*.

King, unlike Davis, has pursued the endogenetic component of his model vigorously and supported a not very popular model of the Earth's planetary development in doing so. Tectonic movement is widely accepted as being subdivided into epeirogeny (broad, relatively uniform uplift of large areas) and orogeny (mountain-building uplift, often exhibiting intense deformation and occurring in linear belts). King added to this widely accepted pair a third, less acclaimed, process called cymatogeny. The basic attributes of cymatogeny are the arching, and sometimes doming, of zones tens or hundreds of miles wide and characterized by great vertical uplift (many thousands of feet), but relatively little rock deformation. Figure 10.7 (King 1967, p. 657) provides a summary of the principal attributes of cymatogeny.

Cymatogeny and pediplanation have been integrated by King in a fashion

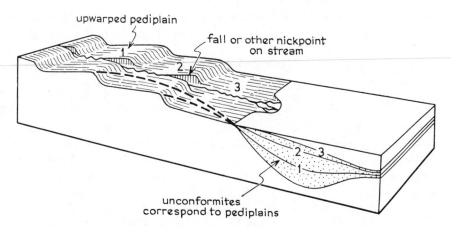

Figure 10.8   An illustration of L. C. King's model for the development of the major landscape features in southern Africa. (From Dury (1969).)

that is summarized in Figure 10.8. The underlying principle here is that as the land mass is eroded sedimentation occurs offshore. The unloading of the land mass causes uplift along the margin of the continent, while the loading of the offshore region causes depression. In turn, this relationship initiates a major scarp in the coastal region. Such a major scarp then retreats inland, exhibiting parallel retreat just like any other scarp or hillslope. However, the surface between each of these scarps is a pediplain and as such it experiences little or no modification as its area is reduced by parallel retreat of the encroaching scarp.

The overall effect of this set of ideas may be summarized briefly. Erosion is cyclical as Davis suggested (but King has also provided a reason why this is so – the previously mentioned loading–unloading relationship). At the continental scale there are massive erosion surfaces that form a huge staircase. Each of these surfaces is a pediplain, time-transgressive in age, but largely unaltered since its formation. Each new erosion cycle starts in the coastal region and advances inland. However, although each new cycle reduces the area of the preceding pediplain by incising into its margins, it does not modify in any meaningful way the nature of the older pediplain surface above the margin that is being consumed. Therefore, high interior surfaces of the various continents are very old, but largely unchanged since their formation.

King has continued to pursue his efforts at geomorphological modeling, venturing far into the realms of plate tectonics and geophysics (King 1983). While King does not deny the existence of epeirogeny and orogeny, he believes that cymatogeny is the dominant mode of crustal movement. The fundamental difference here is that conventional interpretations of plate tectonics, of necessity, involve substantial horizontal movements of the crust. In cymatogeny, vertical movement is dominant and the argument is supported by a belief in an expanding Earth (e.g., King 1983, pp. 113, 122–4).

King's extensive efforts in the realms of landscape evolution and crustal mobility are brought together by his identification of a number of globally distributed erosion surfaces. His most recent summary of these is to be found in King (1983, pp. 177–96). While not all surfaces are found everywhere, King has found evidence scattered across the world of Gondwana planation (Jurassic), Kretacic planation (early to mid-Cretaceous), Moorland planation (late Cretaceous to mid-Cenozoic), Rolling landsurface (Miocene), Widespread landscape (Pliocene), and Youngest cycle (Modern–Quaternary). King (1983, pp. 197–8) also observed that, worldwide, all submarine and sub-aerial surfaces have been dated as Jurassic or younger. He believes that mid-Mesozoic erosion or burial destroyed evidence of all earlier surfaces.

King's geomorphology: strengths and weaknesses

The most obvious strengths of King's model are its comprehensive nature and the forcefulness with which King has presented it. Like Davis and Penck, King presents his model as one that is applicable to all sub-aerial portions of the Earth's surface. With respect to endogenetic forces, King's ideas are even more complete than Penck's, but at odds with Penck's noncyclical perspective. In sum, King provides geomorphologists with a truly comprehensive, large-scale model of geomorphic evolution and its endogenetic origins (at least as far as modern tectonics is capable of taking it). Given such an ambitious undertaking, there is much scope for carping!

A primary concern within the geomorphic component of King's model is his pre-eminent ranking of process, versus a total absence of process measurement in his research. This cannot be presented as wholly invalidating his model because there are certainly few relevant process data at the scale at which King has worked. However, there are now some preliminary studies on the massive scarps that appear to rim not only most of southern Africa but many other continental margins as well. Ollier (1984) made a preliminary survey of possible formative mechanisms for major, continental scarps and, while not drawing firm conclusions, tended to favor marginal uplift following the breakup of Pangea, plus subsequent scarp retreat.

As King has invoked identical behavior of slopes small and large, it does not seem unreasonable to suggest that he might have undertaken field measurement at the smaller scales. Current research in slope geomorphology has yet to identify any pattern of development as universal, although process research on surface flow has certainly reached a point where King's claim for the widespread presence of laminar flow on pediment surfaces seems most unlikely to be valid.

Cyclicity has an innate appeal to geomorphologists and geologists (see Higgins (1975, pp. 14–15) for some interesting comments on this topic). It is certainly possible to marshal evidence in its favor (e.g., Melhorn & Edgar 1975), but at the scale under discussion dating is usually sufficiently imprecise that the establishment of true cyclicity is usually impossible. While noting the uncertainty of dating procedures, Summerfield (1984) has shown that King's proposal relating the retreat of scarps along the coastal margins of southern Africa to offshore sedimentation appears erroneous given the available dates from both environments. If interpreted cautiously, King's work can be said to be no better, and no worse, than other attempts to model at this scale.

King has offered no unequivocal evidence to support his concept that ancient pediplain surfaces are essentially unaltered. Again scale is a confounding issue; if pediplains are time-transgressive, just how much variability in landform assemblages should be expected or permitted? What is the appropriate yardstick to apply? In a recent review of the southern African landscape Summerfield (1984) concluded that the available evidence did not favor King's concept of the development of continental-scale erosion surfaces. Again King's lack of field measurement leaves this issue hanging. When trying to establish worldwide correlations these same issues simply become even more contentious. In reading King's work it is hard not to develop the impression that almost any variation that might seem to negate his concepts somehow falls into the realm of what he considers reasonably expectable variability.

King's model of cymatogeny takes him well beyond the bounds of what is normally considered geomorphology and well into the realms of plate tectonics and geophysics. It is fair to note that cymatogeny is clearly a minority opinion, as is the expanding-Earth concept on which it is now superimposed. In contrast to this perspective it is only reasonable to point out that plate tectonics is still an evolving and highly debated issue. Some very eminent geologists remained unconvinced of important aspects of the plate tectonics model, and as recently noted by Saull (1986) geology badly needs some competing hypotheses. Again Summerfield (1984), in his review of the issue, found little present-day evidence to support this element of King's model.

It is very unfortunate that King is out of step with the times in geomorphology. His ideas are grandiose, but they are worthy of great attention and a determined effort to disprove them (remember as scientists we can only disprove!). It is true that his writings are somewhat hectoring in tenor but, given the difficulty of constructing a framework of this magnitude whilst simultaneously trying to attract the attention of a rather indifferent audience, he would surely create such a style if it were not inbred. Judgment on King's model remains to be made because, as Young (1972, p. 38) wrote, "The value of King's work is in setting up hypotheses for testing; they are not proven".

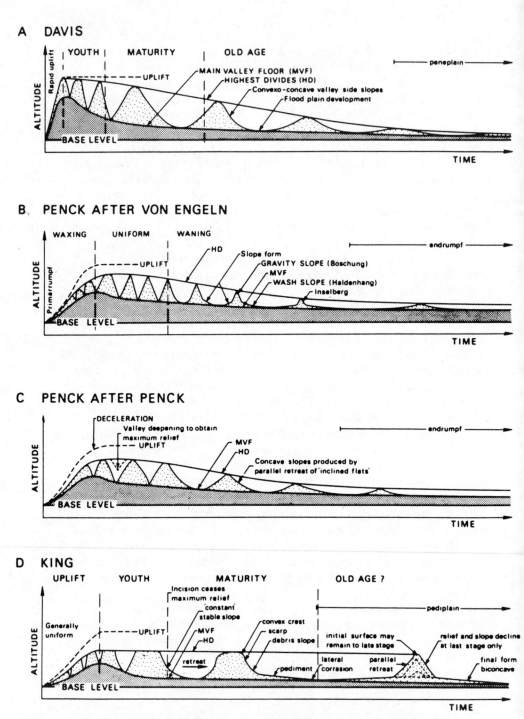

Figure 10.9 A schematic comparison of landscape development as envisaged by Davis, Penck, and King. (From Thornes & Brunsden (1977).)

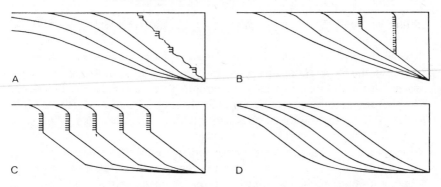

Figure 10.10 A schematic comparison of slope profile development as envisaged by Davis, Penck, and King: (a) slope decline; (b) slope replacement; (c) and (d) parallel retreat, in (c) with a free face and in (d) without a free face. (From Young (1972).)

Some comparative comments

True or false, myth or reality, the three models summarized in this chapter represent the framework and options within which landscape evolution has been considered in the English-speaking geomorphological community since about 1900. As landscape evolution concepts, they are summarized in Figure 10.9; as models of slope evolution, in Figure 10.10.

Amongst the three, Davis's model was first and has been most influential. It is unfortunate, but probably true to say, that the misconceived version of Penck's model served rather more as an alternative to Davis's than did the real thing. By the time the true version of Penck's ideas had been appreciated, the topic itself had lost much of its appeal. In many ways King's model serves both as a link to the past and as a beginning of the future. Its link to the past is obvious as King has overtly drawn on both Davis and Penck; while the present is represented by his own pre-eminence as a student of this scale of research, as well as by his somewhat controversial wedding of his erosion model to plate tectonics. The future is embraced because King's model is essentially untested and therefore, if geomorphologists decide as a group to return to this research scale, testing King's model will be an obvious starting point.

Examination of individual components of each model has already been undertaken and only one or two important generalizations will be added here. While these three models are often considered to belong to classical geomorphology, meaning that they lack a quantitative methodology, there is a fundamental difference between Penck's ideas and those of Davis and King. Davis and King present time-dependent, cyclical models. This contrasts sharply with Penck's model, which is neither time-dependent nor cyclical. In fact, the very essence of Penck's concept is that there is continual adjustment of erosion in response to ongoing, but variable, rates of crustal mobility (including stability).

Ahnert (1970, p. 262) made some calculations on the rate at which erosion could be expected to come into equilibrium with uplift rate. He concluded that uplift would have to remain constant for tens of millions of years before equilibrium could be reached. Therefore, if Ahnert is roughly correct in his computations, Penck's idea concerning relationships between uplift and erosion are unlikely to be correct.

Relationships between uplift rates and denudation (using the normal, non-Penckian definition) rates are inherently vital to all three models. Schumm (1963b) surveyed published rates for both processes. He reported average denudation rates of 0.1–0.3 ft per 1,000 years (with average maximum rates of about 3 ft per 1,000 years). Rates for modern orogeny were about 25 ft per 1,000 years. This discrepancy has often been used to suggest that Davis's simplification of rapid uplift, followed by erosion, is acceptable. In fact, the denudation estimates should be viewed cautiously as they were taken from studies using river sediment loads as the output. Large rivers interact primarily with their floodplain deposits and, be it overestimate or underestimate, this does not necessarily tell much about overall surface lowering within basins (e.g., Trimble 1977).

Schumm (1963b, p. H9) went on to make some simple calculations concerning time required for peneplanation (note that he urged a suitable degree of caution in accepting them). His calculations suggested that peneplanation of an area with about 5,000 ft of relief could occur in as little as 10 million years at the maximum rate of denudation, but would take 110 million years at the average rate. He believed this made peneplanation a viable concept in at least some contexts. Ahnert (1970) made similar calculations, concluding that in mid-latitude environments without a pronounced seasonality of rainfall it would take 22 million years without crustal movement, and 37 million years with isostatic compensation, to reduce relief of an area to 1% of its initial value. He went on to conclude that peneplanation is unlikely under such constraints.

Modern ideas on slope evolution (Young 1972, Carson & Kirkby 1972, Selby 1982, Abrahams 1986) are much advanced over those expressed in the three large-scale models. However, they present models of greater complexity, rather than providing categorical support for, or rejection of, any one of the three classical models. In any event, slope research is not being pursued at a scale that is compatible with the needs or objectives expressed in this chapter.

In short, landscape evolution on the regional scale is an abandoned theme in modern geomorphological research, at least by anything more than a handful of individuals. Nevertheless, concepts developed earlier in pursuit of the goal of understanding regional landscape evolution still pervade the discipline. Primarily, this seems to be represented by widespread willingness to revert uncritically to Davisian perspectives. This is often defended on the grounds that the Davisian approach is an "acceptable introductory teaching vehicle". Such a claim is really subject to two restrictions. First, it is true only if the intent is to teach regional geomorphology at the scale at which Davis worked. Second, even if the first criterion is met, it is critical to make it clear that Davis's model represents only one option and is not *the way* in which landscapes develop. One individual, among the handful of those who have continued to pursue regional geomorphology, is John T. Hack. He has proposed a radical alternative to the time-dependent and cyclical concepts of Davis and King, and it is Hack's ideas that form the focus of the next chapter.

11 A time-independent model

The models of Davis, Penck, and King differ in some fundamental ways, not least of which is the manner in which time is reflected in landscapes; nevertheless in each of these models landscapes bear imprints of times past. Given the focal and pervasive roles of time and space in human thought, it is hardly surprising to find time dominant in some models of landscape development; nor would it be surprising to find space fulfilling this same role. In fact, space is the fundamental ingredient in an alternative approach to landscape modeling.

In viewing landscapes from a temporal perspective it is readily accepted that some portion of variability stems from spatial inputs. A simple example of this would be recognition that limestone and granite rock types will produce different landscapes purely as the result of the inherent differences in the rocks themselves. When examining landscapes from a temporal perspective, this kind of spatial variability is seen as contamination or noise in the signal. When landscapes are viewed from a spatial perspective it is time or the temporal signal that represents contamination. Whatever the central theme of a model the presence of contamination does not entirely invalidate it.

In a time-independent landscape model the creator attempts to develop a theoretical structure in which landscape differences are assigned not to variations in age (whether it be measured absolutely or relatively) but to some spatial input that varies. In Penck's case the central theme was the spatial variability of crustal movement, but the past record of crustal mobility was still to be seen in the landscape. Other spatial themes exist around which to build a model and the influence of past events may not be viewed as present at all. One very appealing approach is to assume some sort of equilibrium between contemporary surficial processes and the surface upon which they are acting. If such a balance or equilibrium exists there is no need to appeal to past events (with the attendant uncertainty) to explain present-day landforms. Two problems immediately present themselves: (1) Which of the available equilibrium concepts should be selected? (2) How much contamination is acceptable before the model is considered ill-founded? The latter question should be asked of any model, but quickly comes to the forefront in this particular context because time, the main source of contamination, is also the focus of the better-known models in geomorphology.

Models founded on spatial variability (or time-independent models) have not enjoyed the ready acceptance of time-dependent models. There are probably a number of factors that account for the discrepancy. When regional-scale geomorphology was popular, it was not very quantitative. This situation was matched by a vigorous champion of a nonquantitative, time-dependent model, namely Davis. The other great American master of geomorphology at the time, Grove Karl Gilbert, pursued an approach to geomorphology that was eminently suited to the development of a time-independent model. However, while Gilbert was certainly Davis's intellectual equal and his work quantitative, he made little or no effort to present a comprehensive model of landscape development to the discipline. Furthermore, Davis's time-dependent approach (derived ultimately from biology) was certainly more in tune with the times than was Gilbert's approach, which was founded on physics and engineering (Baker & Pyne 1978, p. 98).

As noted in the previous chapter, in recent decades the reappraisal of Davis's ideas has been reflected or matched by an increase in the role of process geomorphology. One sidelight on this trend has been a renewed interest in the work of G. K. Gilbert. A specific product of this combination has been John T. Hack's introduction of a time-independent landscape model for the Appalachians – the very region that inspired Davis's model. Hack has traced his ideas back to G. K. Gilbert while acknowledging the role played by A. N. Strahler. In examining Hack's model the "Scientific context" section will comprise an undisguised attempt to review, briefly, some of the important contributions made by Gilbert.

John T. Hack

Scientific context

John Hack has produced a number of carefully crafted geomorphological papers focused on the Appalachians (e.g., Hack 1960, 1975, 1980, 1982). His primary methodological contribution has been to champion the concept of dynamic equilibrium. As he himself noted (Hack 1960, p. 81) his emphasis on dynamic equilibrium was a conscious and deliberate attempt to view the Appalachians (whose development had long been dominated by Davis's model) in a new light.

Dynamic equilibrium is founded on the idea that contemporary landforms are in balance with present-day processes. If such a condition exists the landscape does not need to be interpreted in a historical, or time-dependent, manner but only in terms of existing relationships between available forces and variable surface resistances. While Hack's papers may be cited as the stimulus for present-day attention to dynamic equilibrium, Hack (1960, p. 81) pointed out that his ideas are drawn directly from those of G. K. Gilbert.

G. K. Gilbert, a contemporary of Davis, received great personal recognition during his lifetime, but his scientific ideas lay fallow for many years. His failure to attract a large scientific following comparable to that of Davis seems fairly easy to explain and rests on the following: (1) Gilbert was not particularly interested in a historical approach to geomorphology (this set him very much apart from the orthodox geomorphology of his time); (2) he worked for the U.S. Geological Survey and not at a university (tending to minimize his interaction with academics and students); (3) a sizable portion of his career was consumed by administration (disrupting some of his classic work and obviously robbing him of many productive years); (4) he made no attempt to construct a comprehensive model from his ideas (effectively ceding this territory to Davis, despite the fact that Gilbert's ideas clearly differed sharply); (5) Gilbert appears to have been very unassuming, both personally and professionally (putting him very much at a disadvantage at a time when Davis was actively attracting the limelight). These traits (distilled from Baker & Pyne (1978)) are certainly sufficient to account for the differences in the way in which two important sets of ideas have made an impact on the discipline.

The contrast in intellectual foundations, Davis's allegiance to biology and Gilbert's to physics and engineering, undoubtedly accounted for the lack of temporal domination in Gilbert's thinking. His perspective was probably also reinforced by his work in the western U.S.A. where the starkness of semi-arid landscapes tends to reinforce notions of their extreme youthfulness. Gilbert's methodological contributions generally are part and parcel of his research papers, of which the Henry Mountains report (Gilbert, 1877), the Lake Bonneville monograph (Gilbert, 1890), and the fluvial transport paper (Gilbert, 1914) are probably the most noteworthy; nevertheless he did publish one notable paper specifically on methodology (Gilbert, 1886).

In his "The inculcation of scientific method by example", Gilbert (1886) presented his belief in the "working hypothesis", an approach that is usually associated with Chamberlin's (1897) later presentation of it. As noted by Baker & Pyne (1978) Gilbert's method is pervaded by the use of analogies. While the analogy was his primary method of generating hypotheses, he wrote most forcefully against a researcher supporting a single, or favored, hypothesis. Gilbert wanted many hypotheses created and then exposed to rigorous testing. He specifically rejected the gathering of support for a single hypothesis (a trait he assigned to "theorists"), and championed the approach of the "investigator", an individual who "seeks diligently for the facts which may overthrow his tentative theory" (Gilbert, 1886, p. 286). Gilbert's choice of terminology leaves something to be desired but the thrust of the argument is unimpeachable.

Despite the absence of a comprehensive landscape evolution model, Gilbert did present a number of important generalizations. These were forwarded as "laws" and are listed by Higgins (1975, p. 5): (1) "law of uniform slope"; (2) "law of structure"; (3) "law of divides"; (4) "tendency to equality of action, or to the establishment of dynamic equilibrium". All of these are to be found in Gilbert's (1877) Henry Mountains report.

The law of uniform slope (Gilbert 1877, p. 115) establishes that the rate of erosion increases with increasing slope angle. Furthermore, the rate of increase is greater than a linear one. The law of structure (Gilbert 1877, pp. 115–16) represents identification of the differential erosion experienced by resistant and unresistant materials. In addition, Gilbert noted that "equality of action" could be established between hard and soft materials through adjustment of slope angles. In other words, the rate of erosion could be the same for both a hard and a soft material by virtue of the hard material being a topographic high and/or steep, and the soft material being a topographic low and/or gentle in slope. The law of divides stated (Gilbert 1877, pp. 116–17) that stream gradients increase as the watershed (divide) is approached and, furthermore, that mountains are steepest at their crests. The law of divides subsequently became the "law of increasing aclivity", and Gilbert (1884, p. 75) later created the "law of interdependence of parts" which subsumed both (2) and (4) above.

The equality of action identified by Gilbert in his law of structure was expressed at greater length a little later in his report on the Henry Mountains. The fact that it is a fully developed and sophisticated grasp of dynamic equilibrium is beyond doubt, as the following quotation shows (Gilbert 1877, pp. 123–4):

> The tendency to equality of action, or to the establishment of a dynamic equilibrium, has already been pointed out in the discussion of the principles of erosion and of sculpture, but one of its most important results has not been noticed.
>
> Of the main conditions which determine the rate of erosion, namely, quantity of running water, vegetation, texture of rock, and declivity, only the last is reciprocally determined by rate of erosion. Declivity originates in upheaval, or in the displacements of the earth's crust by which mountains and continents are formed; but it receives its distribution in detail in accordance with the laws of erosion. Wherever by reason of change in any of the conditions the erosive agents come to have locally exceptional power, that power is steadily diminished by the reaction of rate of erosion upon declivity. Every slope is a member of a series, receiving the water and the waste of the slope above it, and discharging its own water and waste upon the slope below. If one member of the series is eroded with exceptional rapidity, two things immediately result: first, the member above has its level of discharge lowered, and its rate of erosion is

thereby increased; and second, the member below, being clogged by an exceptional load of detritus, has its rate of erosion diminished. The acceleration above and the retardation below, diminish the declivity of the member in which the disturbance originated; and as the declivity is reduced the rate of erosion is likewise reduced. But the effect does not stop here. The disturbance which has been transferred from one member of the series to the two which adjoin it, is by them transmitted to others, and does not cease until it has reached the confines of the drainage basin. For in each basin all lines of drainage unite in a main line, and a disturbance upon any line is communicated through it to the main line and thence to every tributary. And as any member of the system may influence all the others, so each member is influenced by every other. There is an interdependence throughout the system.

It is from this statement of dynamic equilibrium and the related cluster of physically based concepts that Hack drew his representation of dynamic equilibrium. For those who wish to delve more fully into Gilbert's contribution to geomorphology the best starting point is Pyne (1980), while Baker & Pyne (1978) and Chorley & Beckinsale (1980) will provide much briefer reviews.

Dynamic equilibrium

The gist of dynamic equilibrium as promoted by Hack is embraced in the following quotations (Hack 1960, pp. 85 and 86 respectively):

An alternative approach to landscape interpretation is through the application of the principle of dynamic equilibrium to spatial relations within the drainage system. It is assumed that within a single erosional system all elements of the topography are mutually adjusted so that they are downwasting at the same rate. The forms and processes are in a steady state of balance and may be considered as time independent. Differences and characteristics of form are therefore explainable in terms of spatial relations in which geologic patterns are the primary consideration rather than in terms of a particular theoretical evolutionary development such as Davis envisaged.

The concept requires a state of balance between opposing forces such that they operate at equal rates and their effects cancel each other to produce a steady state, in which energy is continually entering and leaving the system. The opposing forces might be of various kinds. For example, an alluvial fan would be in dynamic equilibrium if the debris shed from the mountain behind it were deposited on the fan at exactly the same rate as it was removed by erosion from the surface of the fan itself. Similarly a slope would be in equilibrium if the material washed down the face and removed from its summit were exactly balanced by erosion at the foot.

If dynamic equilibrium is compared to a steady state and dynamic metastable equilibrium (Ch. 6), in each instance there are similarities, but at least one important distinction. A steady state exhibits fluctuations about an unchanging mean value; this may be contrasted to the fluctuations in dynamic equilibrium, which occur about a distinctly trending mean. However, regardless of the length of time in which the researcher is interested, if only a small segment of it is sampled it may be difficult or impossible to distinguish between a steady state and dynamic equilibrium. Dynamic metastable equilibrium differs from dynamic equilibrium because the former includes large, abrupt changes as well as the smaller fluctuations about the overall trend which are common to both concepts. These larger jumps are now commonly referred to as thresholds in geomorphology (Ch. 6) and have been focal in Schumm's (e.g., Schumm 1979) model of landscape development. Again, if only a small segment of the timespan of interest is sampled it may be difficult, if not impossible, to distinguish between dynamic equilibrium and metastable dynamic equilibrium. This failure is indeed probable if the sampling period happens to occur between thresholds.

An important distinction between Hack's promotion of dynamic equilibrium and the models of the previous chapter is that Hack does not believe that dynamic equilibrium represents a comprehensive model (Hack 1975, p. 88). It is a perspective or viewpoint that may be usefully applied over the entire spatial range that interests geomorphologists, from the very small to the very large (Hack 1960, pp. 86–7), but it does not require total rejection of geomorphic evolution – this means that temporal change may be invoked.

A general example presented by Hack (1960, p. 93) is a contrast between a resistant rock type and a much less resistant one. Both rocks are assumed to be downwasting at the same rate – this is the fundamental dynamic equilibrium concept. However, it requires more erosive energy to reduce a resistant material than a less resistant one. Therefore, the resistant material has higher relief and steeper slopes (thereby exhibiting higher potential and kinetic energy) than the surface of the less resistant rock. It is also possible, although not essential, that they may exhibit different weathering regimes; for example, high relief and steep slopes may mean domination by physical processes, while the more subdued, less resistant surface is primarily reduced by chemical weathering. As an aside it is worth noting that this suggested contrast (Hack 1960, p. 93) is not generally supported by more recent research in alpine areas.

Such a general example may be made more specific (Hack 1960, p. 93). In many Appalachian valleys there are features that resemble alluvial terraces;

indeed, this would be a time-dependent explanation of their origin. However, such areas normally occur in valley bottoms where the abutting ridges have extensive outcrops of resistant rocks. Boulders and cobbles are brought down by the tributary streams to a trunk stream that is adjusted to the surrounding softer materials in the main valley. Lacking competence (the competence of a stream is the largest fragment it can transport) to handle this coarse material, the main stream deposits the boulders and cobbles and shifts laterally. This produces a valley bottom that is armored with more resistant coarse material. Eventually such areas stand above the rest of the valley bottom, looking just like a fluvial terrace, but lacking an origin derived from rejuvenation or a similar time-dependent concept. These pediments (so-called by Hack 1960, p. 93) are the result of the interdependence of all parts of the landscape, mutual adjustment being central to dynamic equilibrium (see the quotation from Gilbert above). However, it is apparent that in this instance the pediment surfaces cannot emerge above the level of the rest of the valley floor without an intervening period of disequilibrium.

Under the specifications of dynamic equilibrium the landscape is expected to downwaste without obvious changes in form, unless there is a change in energy inputs or surface resistance. Hack (1975, pp. 94–100) discussed some possible outcomes of changing regional base level, and it is easy to envisage that erosion may expose a rock type of differing resistance from the one removed. In either of these circumstances, surface form would be expected to change without violating the dynamic equilibrium concept (obviously the boulders and cobbles being deposited on the valley bottom in the above example modify surface resistance and therefore the ensuing period of change is one of disequilibrium, but falls within the overall concept of dynamic equilibrium).

In relating dynamic equilibrium to existing models, Hack raised two interesting points. First, it is possible to make an argument parallel to Penck's belief that rate of uplift and rate of erosion are matched (Hack 1960, p. 86). However, Hack couches this in terms of rate of uplift plus surface resistance being matched by erosion rate. Indeed rapid uplift would require high elevations and large relative relief to produce the necessary rate of erosion. Second, widespread occurrence of "maturely dissected peneplains" suggests that they, and not the peneplain, are the equilibrium form derived from a long period of erosion uninterrupted by tectonic movement (Hack 1960, p. 89). In fact, "ridge and ravine" topography (Hack's suggested terminology to bypass the genetic connotations of "maturely dissected peneplain") appears to exhibit slope forms that are more akin to those postulated by Gilbert (1909) than with those that would occur with dissection of a peneplain. In two more recent papers, Hack (1980, 1982) has sustained and amplified his argument using analysis of available uplift and erosion rates. While he was able to show that there is broad agreement between uplift and rock resistance on one hand and erosion plus relief on the other, the data base is still sufficiently sketchy to mandate caution in its acceptance and interpretation.

Dynamic equilibrium: strengths and weaknesses

As noted earlier, an important principle in science is acceptance, not of the simplest answer, but of the simplest explanation that may be matched successfully with existing theory. A logical extension of this principle is to create theory that is as simple as possible. To this principle may be added the observation that the past is, at best, always only partially known and in many instances largely unknown. Integration of these ideas suggests that a body of theory in which the role of past events is minimized and present processes are assumed to be in balance should be a very appealing one intellectually. Clearly, dynamic equilibrium presents this combination and, if valid, provides not only a well-defined set of rules but also a set that offers a very positive role to process geomorphology.

The most fundamental issues involved are the spatial and temporal scales being considered because there are no questions in geomorphology that can be meaningfully posed or answered without specification of applicable scales. In the case at hand this is essential because it will do much to clarify the seeming dilemma between time-dependent and time-independent behavior. As Howard (1965, p. 302) pointed out, the influence of past processes is proportional to their intensity, but inversely proportional to the time elapsed since they occurred. This observation at least establishes a principle upon which to incorporate or eliminate time as a factor in most circumstances; although it is inevitable that there will be some (many?) situations where unknown or unrecognized events will have been important. However, wherever and whenever time is reduced in an explanation, spatial variability must be expanded to fill the explanatory void.

As is usual with such problems, the end members of the time-dependent/time-independent distribution appear to be fairly easy to identify. Time-independent behavior is unlikely to be invoked in an explanation of the geomorphology in and around Mount St Helens in the near future. Conversely, the seasonal rhythm of Boomer Beach at La Jolla, California (Shelton 1966, pp. 186–7) seems to exhibit some sort of equilibrium behavior. Furthermore, single landforms may be cited as exhibiting both types of behavior simultaneously; for example, an underfit stream exhibits time-dependent behavior with respect to its valley (Dury, 1964, 1965), but time-independent behavior with respect to its channel form – if indeed it exhibits the discharge-to-channel relationships that most streams do (Leopold & Maddock 1953). The problem is that examples are offset by a multitude of forms exhibiting unknown or uncertain combinations of behavior. It is fair to point out that most other models face similar difficulties.

The second major issue confronting dynamic equilibrium is twofold: (1) Is equilibrium in the general sense an applicable concept? (2) If equilibrium in general is applicable, is dynamic equilibrium the best version to use? The first issue has been addressed already in Chapter 5 where it was noted that geomorphologists have generally treated landforms as either equilibrium or relaxation forms. The stringent constraints associated with equilibrium or characteristic-form modeling have already been noted (Ch. 5). Clearly, the existence of relaxation models indicates that many geomorphologists have rejected equilibria approaches altogether. A specific example of disequilibrium is Church & Ryder's (1972) concept of "paraglaciation", wherein a recently deglaciated area experiences a period of high sediment yield as glacial depositional forms adjust to their new environment.

If an equilibrium approach is chosen, selection of which type would seem to require detailed, and lengthy, records. It seems essential to select an equilibrium that reflects an open system, in which case the most obvious choices would appear to be between steady state, dynamic equilibrium, and dynamic metastable equilibrium. Despite what seems to be some confusing use of the terms "steady state" and "dynamic equilibrium" by Hack, it seems logical to assume that a steady state (i.e., no trend in the mean with time) should be rare. When uplift is present it seems likely that it will exceed erosion rates (e.g., Schumm, 1963b, Ahnert 1970). When uplift is absent, there is likely to be an overall lowering of the landscape. In addition, there are many situations where rock characteristics alone are sufficient to establish ongoing divergence in relief (e.g., Bull 1975b). An example of the latter suggested by Bull is the ongoing increase in relative relief as a lava-filled valley produces first inverted relief by becoming a lava-capped ridge, and then continues to exhibit non-steady-state behavior as the relative relief between valley bottom and ridge crest increases.

The distinction between steady state and dynamic equilibrium requires identification of contrasting behavior of mean values over a specified period. Distinction between dynamic equilibrium and metastable dynamic equilibrium requires identification of two different magnitudes of oscillation about similarly trending means. Again it is perfectly feasible to envisage situations where an abrupt and/or large-scale change in a pattern may be discerned; but again it is equally easy to see the possibility for endless debate as to what constitutes the difference between a big (but "normal") oscillation versus a threshold. Such a distinction would have to be made if dynamic and dynamic metastable equilibria were to be distinguished from each other. The need for extensive data bases upon which to make the distinction is self-evident – as is the paucity of such records.

In conclusion, dynamic equilibrium offers an important intellectual framework within which to place landscape evolution. However, at present it stands as a conceptual option, rather than as a fully tested, corroborated hypothesis. Howard (1965) suggested that a systems approach permits resolution of many of the seeming conflicts between time-dependent and time-independent models. The critical concept here is that systems move toward equilibrium at a speed proportional to their distance from it, i.e., those which are far from equilibrium move rapidly toward it (i.e., they may be viewed as time-dependent) while those that are very close to equilibrium move toward it very slowly (they may be seen as time-independent). Such a viewpoint unites both schemes, as all landforms may be seen as tending toward equilibrium.

The problem between envisaging landforms as tending toward equilibrium, as opposed to actually exhibiting some sort of equilibrium, is that the researcher loses a touchstone or metric and replaces it with only a fuzzy concept which lacks the potential for falsification. This may, in fact, be largely a semantic point given the normal level of resolution available in geomorphology. Even if research invoking dynamic equilibrium presently lacks the necessary precision to establish landscapes as truly time-independent, it serves two useful purposes: (1) it serves as a focus to remind us that time and space (and not time alone) are the focal themes in geomorphology; (2) it emphasizes the integrated nature of the landscape. The concept of landscape integration is not tied exclusively to that of dynamic equilibrium and is most strongly highlighted in a systems approach – the focus of the next chapter.

12 Systems modeling

Investigation of geomorphology using systems modeling is one of several fundamental contributions promoted by Strahler (1950, 1952a) during the 1950s. Melton (1958a) is an illuminating example of early work pursued in this vein, while Chorley (1962) is usually cited as the initial attempt to link systems methodology in geomorphology definitively to its supposed origin – the general systems theory of von Bertalanffy (1950, 1956, 1962). The reason for qualifying the notion that systems in geomorphology are derived from general systems theory is only intended to highlight the fact that some geomorphologists had informally recognized the worth of approaching geomorphic research from a "systems-like perspective" much earlier. Gilbert's (1886, pp. 286–7) description of geomorphology in terms of a "plexus", as well as his use of "system" (Gilbert 1877, p. 124), clearly illustrates that he worked from a systems viewpoint, albeit undefined formally. In fact, Smalley & Vita-Finzi (1969) have shown that the systems concepts generally found in contemporary geomorphic literature are a confused combination of an older terminology derived from thermodynamics and a newer (and disparate) one derived from von Bertalanffy.

Systems modeling swept through geomorphology as one facet of quantification during the 1960s. Indeed, it is perhaps fair to compare the relationship between systems modeling and process geomorphology with that of Davisian geomorphology and denudation chronology. In each case the relationship is not essential, but became overwhelmingly the norm because of obvious compatibility between the two. Chorley has provided much of the sustained effort to promote systems modeling in geomorphology. Chorley & Kennedy (1971) was an extremely innovative text in geomorphology when it was originally published; today it remains an informative introduction and review of basic concepts in systems modeling as applied to geomorphology. Ideas, and particularly techniques, have continued to grow, with Bennett & Chorley (1978) and Huggett (1985) offering much more sophisticated mathematical treatments of the topic.

The organizational characteristics of systems modeling, plus the feasibility of presenting this in conceptual as well as quantitative terms, offers great heuristic (teaching) advantages. As a result system models have filtered down into introductory texts very widely; for example, Dury (1981),

King (1980), Strahler & Strahler (1973), and at a more advanced level Chorley *et al.* (1984). Consequently, although there are some important intellectual reservations about systems modeling, it may be viewed as the prevailing orthodoxy in both research and teaching of process geomorphology.

Here treatment of systems will be limited to a conceptual outline of what is involved and possible. It will begin with some important definitions, followed by a selection of classification schemes for system models and a discussion of input types that have been used in systems modeling. The review will then move on to an examination of canonical structures (a common way of depicting system models), followed by evaluation of some canonical structures specifically for their geomorphic content. All of these steps are integrated in a brief study of Melton's (1958a) construction of a system model of drainage-basin controls. As a final step the strengths and weaknesses of systems modeling will be reviewed.

Systems and models

Definitions

One of the most frequently cited definitions of a system is (Hall & Fagen 1956, p. 18):

> A system is a set of objects together with relationships between the objects and between their attributes.

These authors described "objects" as the physical (anything from an atom to a galaxy) and/or abstract (e.g., mathematical variables, equations, rules, laws, processes) parts or components of a system; "attributes" as the physical properties of the objects; "relationships" as those things that "tie the system together" (Hall & Fagen 1956, p. 18). An important factor in the last instance is to appreciate that it is up to the individual researcher to choose what are important and unimportant relationships. In geomorphology specifically, it is easy to appreciate that in most instances objects are landform dimensions of one scale or another, relationships are commonly geomorphic processes, and attributes are often such things as rock/soil porosity or drainage density depending on the scale of the research.

The distinction between a system and a model is important because they are not precisely the same thing. A system is essentially an abstract concept being "any set of interrelated or interconnected elements, ... assumed to exist in the real world" (Strahler 1980, p. 1). A model, regardless of how complex, is recognized to be a simplification of reality and will always be partial and consequently incorrect to some degree. Therefore, a true system is something that everyone may agree upon, or believe in, but can never really be known fully. Conversely, a model may be known fully and it may be as simple as its creator wishes, but it must always be remembered that it is incomplete. One byproduct of a model being a man-made artifact is that it can be created for any number of specific purposes using a considerable variety of techniques. As a result, it is important to know the specific purpose of a model because there is no such thing as a universally appropriate one. In this chapter the terms "system" and "system model" will be used to distinguish between the two.

Classification of systems and system models

Systems, like any other group, may be subdivided into categories using a number of criteria. Chorley & Kennedy (1971, pp. 2–3) summarized one important classification of systems according to their function. Three important classes emerged:

(1) *Isolated systems* – Such systems are assumed to have boundaries that prevent the import and export of either energy or mass. Aside from the Universe itself, it is hard to imagine any natural system that is truly isolated. However, the assumption is sometimes made as a matter of convenience, or the situation may be closely approximated in a laboratory.

(2) *Closed systems* – A closed system has boundaries that permit energy to be imported and/or exported, but not mass. The Earth itself is close to this condition, and many geomorphic situations are modeled as if they were closed systems.

(3) *Open systems* – In such a system both energy and mass may freely cross the system boundaries in either direction. Obviously, most of the natural world falls into the open system category.

If a system is isolated it can only use up the energy that it contained initially as none can be imported. Energy moves down gradients from high to low; therefore, if it is distributed unevenly initially there will be a tendency toward equalization. As the distribution becomes more even, so the energy gradients become less steep, and transfer and transfer rates within the system decrease. Conceptually, the final stage is reached when energy is distributed equally throughout the system, and therefore there can be no further movement of it.

The scenario described above is an informal description of the tendency toward maximum entropy as developed in physics to describe the situation in isolated systems (see Ch. 6). Entropy is actually the expected logarithmic probability of the states of a thermodynamic system. However, it is widely interpreted as the randomness of organization in a system. Therefore,

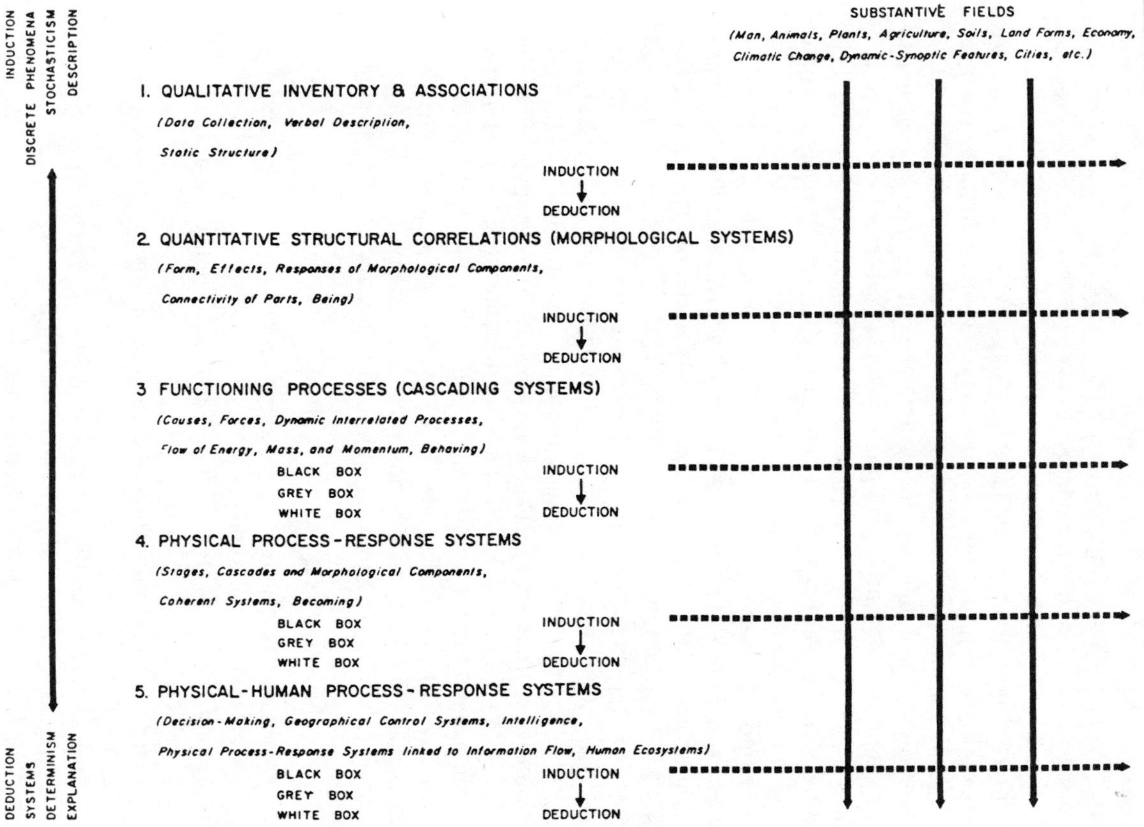

Figure 12.1 A schematic outline of levels of systems modeling within physical geography or geomorphology. (From Terjung (1976).)

maximum entropy occurs when the situation is entirely random, and it also implies that there is no free energy because there are no energy gradients. Conversely, a system exhibiting low entropy is highly organized and exhibits strong energy gradients. In a similar vein "positive entropy" is used to describe the tendency toward maximum entropy, while "negative entropy" (often called negentropy) is used to describe the tendency toward increasing order. It is the free energy in a system that is able to produce work, which, in turn, may produce (land)form change.

A different and very important situation may develop in open systems as a result of their ability to import and export energy and material across their boundaries. The objects and relationships in such a system may, but do not have to, adjust so that there is a steady input and output of energy and mass. Such a situation is produced by a process called "self-regulation" and the result is called a "steady state" (Chorley & Kennedy, 1971, p. 2). This situation is one kind of equilibrium (Ch. 6) and is directly related to the discussion in the previous chapter.

System models may be classified structurally, according to their degree of complexity (Chorley & Kennedy, 1971, p. 3). Such classifications have been made quite frequently, including those by Chorley & Kennedy (1971, pp. 3–10), Terjung (1976, pp. 202–20), Strahler (1980, pp. 3–26), and Thornes & Ferguson (1981). In general terms the classifications are similar, but there are some important differences in definitions. The presentation here will follow the general outline presented by Strahler (1980), while using Chorley & Kennedy (1971) and Terjung (1976) for amplification; the scheme used by Thornes & Ferguson (1981), while of considerable interest, differs sufficiently from the others that it will not be considered here.

Strahler (1980) follows Terjung (1976) by dividing system models into five categories (Fig. 12.1), while Chorley & Kennedy (1971, pp. 3–4) use 11 categories, beginning with the second level listed below. Briefly, Strahler's (1980, p. 2) classes are:

(1) *First level* – Collection of data considered likely to be useful in formulating a functioning system.
(2) *Second level* – Identification and analysis of sets of morphological variables.
(3) *Third level* – Identification and analysis of flow systems of energy and matter.
(4) *Fourth level* – Identification and analysis of process–form systems (integration of data of the second and third levels).
(5) *Fifth level* – Identification and analysis of systems regulated by cybernetic feedback.

Level (1) is extremely crude and includes verbal descriptions and graphing of data. It does not really permit true system models to be created,

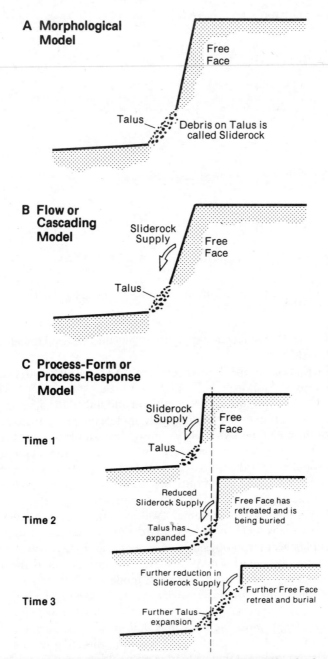

Figure 12.2 A series of increasingly complex system models of a free face (cliff) and talus: (a) a morphological system model; (b) a flow or cascading system model; (c) a process–form or process–response system model.

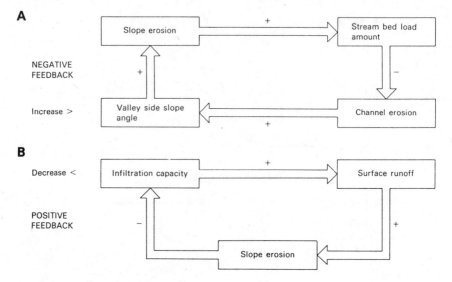

Figure 12.3 Schematic diagrams of (a) a negative feedback loop and (b) a positive feedback loop. (From King (1980).)

although if the data collected are quantified they may be used to help create a system model.

Level (2) represents the lowest level at which a true model may be created. At this level morphological elements (i.e., things that have size, shape, and physical properties) are identified and linked in some meaningful fashion. The variables comprising such models may vary from place to place and/or through time. A simple geomorphic example is a free face with a talus below (Fig. 12.2a). The only inherent attribute of the relationship in the model would be that the talus occurs at the base of the free face. Such a model includes no inference about, nor understanding of, exchange of material and energy between free face and talus.

Level (3) is a flow system (Strahler 1980, p. 10), or a cascading system (Chorley & Kennedy 1971, p. 3). Energy and/or matter pass through the system, linking the morphological components, and may also be stored within it. System boundaries may be natural or artificial, and the system itself may be subdivided into subsystems. A model of the free face and talus would now include the notion that rocks and debris fall off the free face to deliver energy and material (slide rock) to the talus below (Fig. 12.2b). Quite specifically the model implies that slide rock does not jump off the talus and plaster itself to the free face, i.e., the direction of energy and material flows are known. Obviously, this particular example is foolish, but serves to highlight a point that may well not be so obvious in a complex model.

In level (4) process–form links are made (Strahler 1980, p. 11) or the models are of process– response systems (Chorley & Kennedy 1971, p. 3); furthermore, these models represent integration of at least one model of level (2) with at least one of level (3). The most important difference between a process–form model and a flow model (level (3)) is that the concept of feedback is added. Feedback in general terms means that part of the output from one part of the system serves as an input for another part of the system. The concept is usually divided into two opposing relationships, negative and positive feedback (Fig. 12.3). In the case of negative feedback the portion of output that serves as an input elsewhere in the system opposes, or tends to slow, the general direction in which the system is moving. This means that there is a dampening out of this trend. In the case of positive feedback, the output that becomes an input elsewhere tends to enhance, or magnify, the main trend so that it grows rather than diminishes.

Natural systems (with the exception of glacial systems (King 1970)) have a strong tendency to be dominated by negative feedback. This means that any input is likely to be resisted so that the system returns to its original state. Returning to a free face/talus model as an illustration of the various types of system model, it now reaches quite a sophisticated level. Rocks and debris are already known to fall off the free face onto the talus, but up to this time nothing is known about whether or not the relationship changes through time. Now in level (4) the model explicitly deals with this issue (Fig. 12.2c). As rocks fall off the free face they accumulate on the talus at its base. Therefore, the talus will grow in size and gradually bury the free face. In turn, this means that the free face will have a smaller surface area exposed and consequently will supply rock and debris at an ever-decreasing rate to the talus. This means that many relationships in the system model change as the system ages: the free face grows smaller in both absolute and proportional terms while the talus grows larger in both absolute and proportional terms. At the same time the pace of change slows (i.e., negative feedback prevails).

In level (5), the relationships in level (4) prevail, but the entire system is controlled by deliberate human decision making. Strahler (1980, pp. 25–6) actually emphasized cybernetic feedback (requiring a nervous system or electromagnetic communications), while Chorley & Kennedy (1971, pp. 9–10) call such systems "control systems". It is worth emphasizing that the human role in level (5) is conscious or deliberate; inadvertent or accidental modification by human action is incorporated in level (4). This is an important issue in geomorphology because, while many geomorphologists claim to be interested in "natural" systems, man's role in modifying landscapes (even some seemingly wild ones) is now recognized as very widespread. On the other hand some geomorphologists are increasingly involved in engineering or planning issues and as such might pursue geomorphology using level (5) models.

A final classification of models that needs to be noted appears at all levels

Table 12.1   The characteristics of hard and soft systems. (From Agnew (1984, p. 168).)

Hard systems	Soft systems
Well-defined goals	Objectives frequently poorly defined
Well-defined boundaries	Boundaries poorly defined
Clearly established procedures	Decision-taking procedures vague
Quantifiable performances	Difficult to quantify
Clearly structured	Poorly structured
↓	↓
Physical systems?	Human activity systems?

of Terjung's diagram (Fig. 12.1). This is the widely used shorthand terminology of black/gray/white box, a quick and convenient way of characterizing how much is understood about a system. In the case of a black box nothing is known about the internal workings of the system; the only relationship that is known is that a given input results in a predictable output. This might be sufficient in some circumstances; for example, a 10-inch rainfall in 24 h will result in flooding. In a gray box model some of the subsystems are considered, but there is no detailed knowledge or investigation of them. Finally, in a white box situation there is detailed investigation of the entire system, including all of the subsystems and any storage and flows that occur.

Inputs to system models

Now that it has been shown that system models vary considerably in their complexity, it is appropriate to consider what such models require as inputs. In fact, systems modeling does not itself dictate a particular type of input. The full range of possible inputs is perhaps exemplified by the oft-cited subdivision of systems into "soft" and "hard", shown briefly in Table 12.1 and discussed at length in Bennett & Chorley (1978). Soft systems have not generally been used in geomorphology because the hard approach is considered "more scientific". Soft systems are nonquantitative and will not be pursued here; for those interested in pursuing the topic Checkland (1972) and Smyth & Checkland (1976) make a good starting point.

One of the simplest quantitative techniques that permits construction of a system model is correlation. Correlation itself may be subdivided into nonparametric and parametric; the former requires only nominal or ordinal data, while the latter requires interval or ratio data (Siegel 1956, pp. 21–32). As Strahler (1980, p. 6) pointed out, bivariate and multivariate correlation do not distinguish between independent and dependent variables because variance is allocated to both variables. This means that the correlation may be used to measure the strength of the relationship between any two variables, but the researcher must make any cause-and-effect interpretation personally using other considerations.

Melton (1958a) used Kendall's tau (τ), a nonparametric correlation technique, to create a system model of the controls upon drainage-basin development. His inputs were 15 geomorphic, surficial, and climatic variables. Every possible pair was examined, and only those found to be significant at the 92.5% confidence level used. Melton then sorted his variables by looking for isolated groupings (i.e., variables that correlated highly with each other, but not with variables outside the cluster). Such an approach permitted him to identify centers or foci of interaction, but required him to supply his own cause-and-effect interpretation. Melton's paper is discussed in some detail in the following case study to illustrate the basic issues in building a system model founded on a correlation approach.

CASE STUDY – CONSTRUCTING A SIMPLE SYSTEM MODEL

The brief outline provided in this separate section is extracted directly from Melton, M. A. 1958a. Correlation structure of morphometric properties of drainage systems and their controlling agents. *J. Geol.* **66**, 422–60.

The purpose of this short case study is presentation of the techniques and problems involved in building a system model. Melton's paper has three specific advantages in the present context. First, the paper is based on a small data set and the techniques are presented at length. Second, the paper is extensively cross referenced with Miller & Kahn's (1962) text *Statistical analysis in the geological sciences*. Third, the fact that the study was presented exhaustively permits a reader to appreciate the number and nature of the decisions that must be made by the researcher. Given the objective at hand these advantages outweigh the fact that the statistical techniques are now dated. However, for the latter reason, presentation of the statistical procedures is truncated here.

Melton's objectives were investigation of climatic, topographic, and surficial "elements" that might be considered to measure and determine (1) the scale of topography and (2) the scale-free appearance of topography in mature drainage basins. An "element" was defined as "a particular feature of the landscape, climate, flora, etc., that possesses recognized individuality and that can be measured on a simple linear scale" (Melton 1958a, p. 442). Those that are identified as scale-free meet Strahler's dimensionless criteria discussed in this chapter.

Data inputs were 15 variables (Table 12.2) from 59 drainage basins, of which 23 were subjected to a full field study. This immediately establishes one characteristic of a system model, namely that there must be some procedure, or at least decision, for selecting inputs that lie outside the system model itself. Basins were located in Arizona, Colorado, New Mexico, and Utah, and ranged in area from 0.0004 to 5.5 square miles.

Table 12.2   The 15 variables used by Melton (1958a, pp. 446–7) in his drainage-basin study.

(a)   *Morphometric variables*

Valley-side slope (θ) is the angle of the steepest part of a graded valley side; measured largely in the field with an Abney hand level

Drainage density (D) is the ratio of channel length to area drained, in miles; measured from maps

Circularity (C) is a measure of the resemblance of a drainage basin's outline to a circle, given by the ratio of the basin's area to the area of a circle with the same perimeter; measured from maps

Ruggedness number (H) is a scale-free measure of the relative relief of a basin, given by rD, where r is the total basin relief; measured from maps

Ratio of total channel length to basin perimeter (L/P) is a scale-free measure of the relative development of the channel net within a basin's outline; measured from maps

Relative density (F/D^2) is a scale-free measure of the completeness with which the channel net fills the basin outline for a given number of channel segments, given by the ratio of channel frequency to the square of drainage density; measured from maps

Map area of basins of a given order (A_u); mainly fourth-order basins in this study; measured from maps

(b)   *Climatic variables*

Precipitation-effectiveness index (Thornthwaite) (P-E) measures the availability of moisture to plants and is an estimate of the ratio of precipitation to evaporation for an area; obtained from climatic records and field estimates based on flora and elevation

Relative January precipitation intensity (J) is given by the ratio of the mean January precipitation to one-twelfth the mean annual precipitation; data obtained from climatic records

Runoff intensity (q) is a measure of the excess precipitation over infiltration capacity for 5 y, 1 h storms; obtained from Weather Bureau compilations and field infiltration-capacity measurements

(c)   *Surficial variables* (all are field measurements)

Infiltration capacity (f) is the stable, usually minimum, rate at which rain can enter the A_0 soil layer; measured with a sprinkle-plot ring infiltrometer

Wet and dry soil strength (S_w and S_d) is the strength of the soil when thoroughly soaked and when dry, under the impact of a 12 lb shot dropped a standardized distance

Per cent bare area (b) is the percentage of ground surface not covered by vegetation, plant litter, cobbles, bare rock, etc.; measured by counting the number of bare spots under footmarks of a tape measure

Roughness number (M) is the mean total length of rock fragment diameters of ½ inch or more in circles of 1 ft radius, within a drainage basin

---

Given the uncertain nature of the variable distributions, Melton opted to use Kendall's tau (Siegel 1956, pp. 213–23), a nonparametric correlation technique. This means that each pair of variables was correlated separately. The results of this procedure appear in Table 12.3.

Techniques for clustering groups of variables were fairly simple at this time and Melton described his procedure fully, citing Miller & Kahn (1962) as his source. Melton culled his correlation matrix (Table 12.3) by considering only data pairs with an absolute correlation of 0.40 or more at the 92.5%

Table 12.3 A simplified version of Melton's (1958a, pp. 449–50) correlation matrix. Values in italics reached his specified levels of correlation and confidence. Variable values of n (the number of data pairs) in the correlations account for divergence in significant correlation levels.

θ	D	C	H	L/P	F/D²	f	P-E	S_w	S_d	b	M	q	J	A_u
θ ×	+0.02	−0.01	*+0.50*	−0.07	−0.29	*+0.30*	+0.07	−0.16	−0.10	−0.11	0	−0.24	*+0.39*	−0.05
D	×	+0.01	0	−0.03	−0.03	−0.27	*−0.38*	−0.09	+0.04	*+0.65*	+0.19	*+0.43*	*+0.29*	*−0.43*
C		×	−0.16	−0.02	+0.10	−0.18	+0.11	−0.12	+0.06	+0.17	+0.21	+0.28	−0.07	−0.22
H			×	+0.25	*−0.56*	*+0.36*	*+0.13*	−0.17	−0.04	−0.15	+0.11	−0.31	*+0.35*	*+0.18*
L/P				×	−0.32	+0.05	+0.09	+0.09	+0.09	+0.10	*+0.35*	−0.03	+0.04	*+0.51*
F/D²					×	*−0.31*	−0.23	+0.22	−0.09	−0.06	+0.03	+0.25	−0.20	−0.23
f						×	*+0.34*	−0.02	−0.10	*−0.58*	−0.15	*−0.74*	+0.07	*+0.28*
P-E							×	+0.10	+0.09	*−0.68*	*−0.31*	*−0.54*	+0.04	*+0.37*
S_w								×	*+0.50*	−0.17	+0.21	+0.06	−0.18	+0.03
S_d									×	−0.12	+0.20	+0.09	+0.07	−0.13
b										×	*+0.28*	*+0.49*	+0.20	*−0.50*
M											×	+0.19	+0.06	−0.11
q												×	+0.11	*−0.52*
J													×	−0.09
A_u														×

Table 12.4   Basic pairs and description of correlation sets of which they are the nuclei (see text for explanation) identified by Melton (1958a, p. 451).

	Correlation set	Descriptive name
(1)	$\boxed{P\text{-}E \leftrightarrow h} \leftarrow D$	Moisture availability–vegetation set
(2)	$\boxed{f \leftrightarrow q} \leftarrow C$	Infiltration–runoff set
	$\diagdown A_u \leftarrow L/P \leftarrow M$	
(3)	$\boxed{F/D^2 \leftrightarrow H} \leftarrow \theta \leftarrow J$	Relative channel density–ruggedness set
(4)	$\boxed{S_w \leftrightarrow S_d}$	Soil-strength set

confidence level (coefficients meeting the requirement are set in italics in Table 12.3). Clearly, this is a very important decision and one that is not based on any fundamental statistical truth or law. A researcher must simply use personal judgment to determine an appropriate correlation level.

The next step in Melton's procedure was identification of "basic pairs" and "isolated correlation sets". These are variables that correlate more highly with each other than with any other variable. Under such guidelines a basic pair is the minimal subset of an isolated correlation set. Melton obtained four basic pairs (Table 12.4), but no larger isolated correlation sets. "Correlation sets" (Table 12.4) were then created from isolated correlation sets. The difference between an isolated correlation set and a correlation set is that in the former all variables within the set are correlated more highly with each other than with any variable outside the set, while in the second case each variable must only be correlated more highly with a single variable within the set than with any variable outside it. Such a difference may simply be thought of as casting the net wider by lowering admission standards!

The procedure of including variables in a correlation set is iterative; after a variable is added to a group the search is repeated until the group can be extended no further. Melton's correlation sets appear in Table 12.4. Such procedures may be defended as logical, but clearly they are not derived from any fundamental statistical law, nor do they have any geomorphic underpinning.

Melton then recast his correlation sets in a diagrammatic form (Fig. 12.4). Basic pairs and correlation sets were distinguished, and each was enclosed within a distinctive box outline to demonstrate close linkage. In addition, several significant correlations in Table 12.3 between variables that fell into different correlation sets are shown. In fact, another important judgment had been made because a significant correlation between the mean total length of rock fragments (M) and the ratio of total channel length to basin perimeter (L/P) was excluded by Melton as being geomorphically improbable.

The correlation procedure may now be seen for what it is, a precise

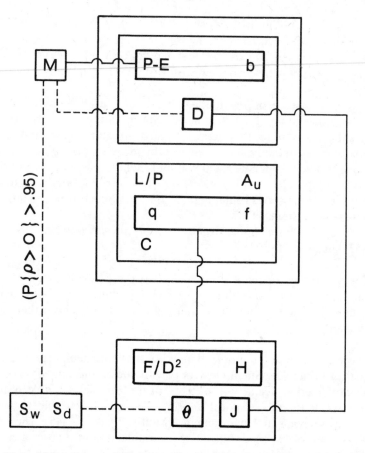

Figure 12.4 Melton's correlation sets shown as a canonical structure. (From Melton (1958a).)

sorting technique that is entirely dependent upon the dictate of the researcher. Initially, the die was cast by variable selection, which clearly constrains what is possible. Now the researcher has constrained the statistical procedure even further by accepting and rejecting correlations on geomorphic grounds. This does not represent anything inherently wrong, but it must be appreciated that such decisions are dictated by the researcher and existing geomorphic theory, and not by the statistical procedure itself.

Melton actually abandoned this particular line of development at this point in the paper and retraced his steps back to the matrix of significant correlations. He then rebuilt his canonical structures using "variable-systems theory", defining a variable system as (Melton 1958a, p. 443):

an abstract set of variables such that (a) each is, in reality, rather highly correlated with every other one; (b) the direction of causality (if any) between each pair of variables is stated; and (c) one or more variables in [the system] may be correlated with variables not in [the system].

The precise nature of the mathematical rules used to determine total clusters, noncontained clusters, and clusters centered on basic pairs is of no great concern here; they represent early attempts to manipulate correlation matrices that have been replaced by more sophisticated techniques. Anyone wishing to pursue them will find them discussed fully in Miller & Kahn (1962, pp. 288–92, 295–9, 304–7). However, it is important to realize that there were a series of mathematical rules applied which identified the members of clusters.

In rebuilding his example from a geomorphic viewpoint Melton (1958a, p. 453) suggested that:

This is largely an inductive process, so there is no reason to restrict variables considered to those for which correlation measures have been obtained, so long as no known natural laws, statistical or otherwise, are violated.

The variable-systems approach incorporated both geomorphological and statistical principles. Small clusters of variables (determined statistically) were explained as negative-, positive-, or no-feedback variable systems (a clear invocation of geomorphic principles). Each variable system was depicted as having an "environment" (Melton 1958a, p. 445). The environment is a set of variables outside the variable system that govern, or are highly correlated with, at least one variable within the system. Rebuilding of the canonical structure following variable-systems theory was founded on the assumption that isolated correlation sets form nuclei of variable systems and that correlation sets contain part of a variable system and its environment. Clearly, this differs from the original correlation approach because some geomorphic hypotheses were inbuilt. Not only were some basic tenets of geomorphology used to interpret the pattern of correlations, Melton even introduced variables that he did not measure to explain and expand upon a number of situations. Only the scale-determining portion of the model will be considered here as an illustration.

Small-component variable systems were presented and discussed individually. One such is the precipitation effectiveness (P-E)/vegetation positive feedback system (Fig. 12.5a). Precipitation effectiveness was (unlike its original presentation) subdivided into regional and local (point) components and relationships between it, vegetation, and bare ground considered. Regional P-E was considered to be controlled by such factors as elevation, latitude, and continentality; while local P-E was considered to be

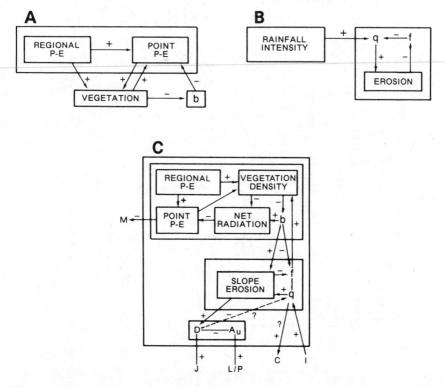

Figure 12.5 Melton's composite variable system model containing climatic and surficial controls of scale of topography (part (c)). Note that regional P-E, point P-E, vegetation, and per cent bare area (b) form one positive feedback system model (part (a)), while slope erosion, infiltration capacity (f), and runoff intensity (q) form another (part (b)). (From Melton (1958a).)

influenced by slope angle and orientation, as well as by surface albedo. Vegetation and local P-E were believed to interact with each other, while regional P-E was considered to influence vegetation without any feedback. Obviously bare ground and vegetation cover were considered to be inversely related.

Another small-scale variable system that was depicted was that between runoff intensity, infiltration capacity, and erosion (Fig. 12.5b). An A horizon is high in organic content and therefore quite permeable. If the A horizon should start to be stripped, for example by an increase in rainfall intensity, the underlying material will contain less organic material and be less permeable; therefore runoff will increase and so produce further erosion. In short, a classic positive feedback cycle will be produced. It will only terminate when a layer of similar or increasing infiltration capacity is encountered during stripping. A final small-scale system forwarded by Melton was an inverse relationship between drainage density (D) and basin area (A_u).

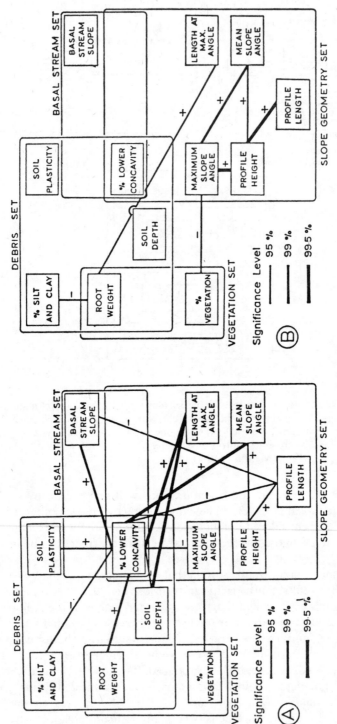

Figure 12.6 Two canonical structures illustrating the contrasts in relationships between slope geometry, debris, and vegetation variables within slopes, when (a) a stream is present at the foot of the slope, and (b) there is no stream at the foot of the slope. Data are from Kennedy's (1965) work on the Charmouthien Limestone, Plateau de Bassigny, Northern France. (From Chorley (1967).)

These three small variable systems were then integrated into a large one (Fig. 12.5c) by Melton. The nature of the relationships was indicated (+, direct; −, inverse), as was the direction of influence (by an arrowhead on the linking line). Where feedback between variables was thought to occur (i.e., there would be an arrowhead at both ends of the link), Melton inserted what might be called a "dummy variable" to clarify the relationship (vegetation density, net radiation, and erosion rate being cases in point). Not all variable links were shown directly; for example, regional P-E was assumed to control D. However, this control is exerted through vegetation and infiltration; as these links already exist within the diagram, no direct link was drawn. It is quite apparent from the diagrams in Figure 12.5 that the final canonical structure was derived from smaller canonical structures, plus insertion of linkages between them. Variables outside the large box represent portions of the environment for this system.

Melton developed his canonical structures derived from variable-systems theory somewhat further, but it does not serve immediate needs to follow these. The important issue is that the inputs to Figure 12.5c do not reflect any sort of "reality" untainted by the researcher's decisions or viewpoint. At every step possible outcomes were highly constrained by initial variable selection. In addition, a mixture of statistical and geomorphic decision making constrained the situation even further at each step by acceptance and rejection of correlations, and finally by the insertion of cause-and-effect inferences that are entirely geomorphic. As with a Davisian diagram so with a canonical structure, the reader sees only what the researcher has chosen to show and not an undistorted representation of reality.

The next step forward from Melton's nonparametric approach is simply to pursue the same topic using parametric correlation. Such models appear frequently in Chorley & Kennedy (1971) and are illustrated in Figure 12.6. In purely illustrative terms there is no difference between nonparametric and parametric models; however, it is obvious that the latter are mathematically superior because they require higher-order measures as inputs. Chorley & Kennedy's text did, in fact, improve the presentation style for such models by distinguishing the various significance levels of the correlations (Fig. 12.6), as well as by labeling the role of each variable symbolically (Fig. 12.7). The links drawn in diagrams such as Figure 12.6 do not necessarily represent flows of energy or matter, but merely statistical correlation. As noted in the case study the direction of influence may be included, but this is a matter of interpretation and not an inherent property of the correlation analysis.

In any scientific endeavor one aim is to be able to predict the behavior of each object under examination. If this is not possible an attempt is usually made to predict the behavior of entire groups of objects with some degree of

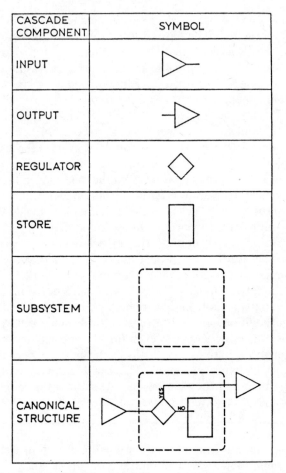

CASCADE COMPONENT	SYMBOL
INPUT	
OUTPUT	
REGULATOR	
STORE	
SUBSYSTEM	
CANONICAL STRUCTURE	

Figure 12.7 Symbols used by Chorley & Kennedy to depict functions in canonical structures. (From Chorley & Kennedy (1971).)

certainty. Statistics come to the fore when group behavior is focal because their primary attribute is that they do not predict the behavior of individuals but can predict group behavior with some known margin of error. Statistics are stochastic, that is, they incorporate randomness, and may be contrasted with deterministic mathematics. In the case of deterministic mathematics the behavior of each individual may be predicted with certainty, or uniquely, using mathematical functions. There is nothing within systems modeling itself that dictates that only stochastic or deterministic mathematics may be used. Therefore, it follows that deterministic system models are preferable to stochastic ones, if they can be created.

Strahler (1952a, 1958, 1980) has emphasized the need for deterministic system modeling using dimensionally correct equations and dimensionless

Table 12.5 Strahler's (1980, pp. 4–6) comprehensive listing of geomorphic variables with appropriate dimensions or units.* Note that the notation system may be expressed in other ways, e.g.,

$$LT^{-1} = \frac{L}{T} = L \cdot \frac{1}{T}$$

Name of variable	Dimensional properties or definition	Dimensional symbol
Class A: Dynamic variables		
Energy (potential, kinetic, internal, radiational)	Force × distance, or mass × velocity squared	ML^2T^{-2}
Work	Same as energy	ML^2T^{-2}
Power	Work or energy per unit of time	ML^2T^{-3}
Rate of energy flow	Same as power	ML^2T^{-3}
Energy flux	Energy flow per unit of cross-sectional area	MT^{-3}
Energy per unit volume		$ML^{-1}T^{-2}$
Acceleration	Distance per unit time, per unit time	LT^{-2}
Force	Mass × acceleration	MLT^{-2}
Stress or pressure intensity (shearing stress)	Force per unit area	$ML^{-1}T^{-2}$
Energy expended per unit surface area (overland flow)	Energy per unit area	MT^{-2}
Momentum	Mass × velocity	MLT^{-1}
Biomass as energy storage	Heat energy equivalent	ML^2T^{-2}
Rate of energy production or storage by photosynthesis	Energy per unit time (power)	ML^2T^{-3}
Rate of energy storage by photosynthesis per unit area	Energy per unit time per unit area	MT^{-3}
Heat	A form of energy	ML^2T^{-2}
Enthalpy	Available energy	ML^2T^{-2}
Class B: Mass-flow variables		
Velocity	Distance per unit time	LT^{-1}
Period		T
Frequency		T^{-1}
Discharge	Volume rate of flow	L^3T^{-1}
Runoff intensity (volumetric)	Volume rate of flow per unit cross section	LT^{-1}
Mass rate of flow	Mass per unit time	MT^{-1}
Mass rate of flow per unit cross-sectional area		$ML^{-2}T^{-1}$
Runoff intensity, rainfall intensity, infiltration capacity	Mass rate of flow per unit area	$ML^{-2}T^{-1}$
Erosion intensity	Mass per unit time per unit area	$ML^{-2}T^{-1}$
Precipitation rate†	Volume per unit time per unit cross section (= depth per unit time)	LT^{-1}
Evapotranspiration rate†	Depth per unit time	LT^{-1}
Infiltration rate†	Depth per unit time	LT^{-1}

Table 12.5 (*cont.*)

Name of variable	Dimensional properties or definition	Dimensional symbol
Bedload movement	Mass per unit time per unit channel width	$ML^{-1}T^{-1}$
Rate of photosynthesis production (net, gross)	Mass per unit time per unit area	$ML^{-2}T^{-1}$
Class C: Geometry variables		
Length, height, depth, distance		L
Area		L^2
Volume		L^3
Number of things	Enumerative	1
Angle, azimuth		1
Length ratio, area ratio		1
Drainage density	Length per unit area	L^{-1}
Stream frequency	Number per unit area	L^{-2}
Circularity, elongation		1
Slope, gradient, hydraulic gradient	Elevation drop per unit distance	1
Curvature	Radians per unit length	L^{-1}
Relative relief, relief ratio		1
Hypsometric integral		1
Bed roughness (absolute)	Length	L
Wavelength, wave height		L
Grain diameter, radius	Length	L
Roundness, sphericity		1
Hydraulic radius	Area/wetted perimeter	L
Fineness, linearity (of bed particles)	Reciprocal of diameter	L^{-1}
Class D: Material-property variables		
Density, bulk density	Mass per unit volume	ML^{-3}
Specific gravity		1
Diffusivity (volumetric)		L^2T^{-1}
Elastic constant		$M^{-1}L^2$
Elastic modulus		$ML^{-1}T^{-2}$
Surface tension		MT^{-2}
Viscosity, absolute or dynamic		$ML^{-1}T^{-1}$
Viscosity kinematic		L^2T^{-1}
Fluid potential		L^2T^2
Weight, specific or unit		$ML^{-2}T^{-2}$
Infiltration capacity	Mass rate of flow per unit cross section	$ML^{-2}T^{-1}$
	Volume rate of flow per unit cross section	LT^{-1}
Resistivity of soil surface (Horton)	Force per unit area	$ML^{-1}T^{-2}$
Erosional proportionality factor (Horton)	Mass rate of erosion per unit area/eroding force per unit area	$L^{-1}T$
Acceleration of gravity		LT^{-2}
Permeability, intrinsic		L^2
Transmissivity (groundwater)		L^2T^{-1}

Table 12.5 (*cont.*)

Name of variable	Dimensional properties or definition	Dimensional symbol
Hydraulic conductivity		LT^{-1}
Specific yield (groundwater)		1
Moisture content of soil	Percentage by weight or volume	1
Porosity	Percent voids	1
Penetrability of beach (Krumbein)	Depth of penetration of dropped ball	L?
Storage coefficient		1
Exchange coefficient in turbulent flow (austausch)	Dynamic viscosity/fluid density	L^2T^{-1}
Suspended sediment concentration	Mass per unit volume	ML^{-3}
Turbidity	Mass of particulates per unit volume	ML^{-3}
Albedo	Percent of energy reflected	1
Cloud cover	Percent of sky covered	1
Reflectivity		1
Infrared emissivity	Percent of blackbody radiation	1
Absorption coefficient	Energy absorption per unit distance	MLT^{-2}
Transmissivity (atmospheric)		1
Thermal quantities (these include dimension (K) of temperature)		
Specific heat	Heat per unit mass	$L^2T^{-2}K^{-1}$
Heat capacity per unit volume		$ML^{-1}T^{-2}K^{-1}$
Heat capacity per unit mass		$L^2T^{-2}K^{-1}$
Temperature		K
Temperature gradient		KL^{-1}
Thermal conductivity		$MLT^{-3}K^{-1}$
Entropy	Energy per unit of temperature	$ML^2T^{-2}K^{-1}$

* Wherever 1 (one) appears as a dimensional symbol in Tables 12.5 and 12.6, Strahler (1980) used the symbol 0 (zero). This use of zero is misleading. "Dimensionless" quantities or numerics are widely used in engineering and physics (angle and relative humidity are examples), as are dimensionless products such as the Reynolds number. When used, these variables are assigned the dimension 1 (one), not 0 (zero); examination of any basic text such as Isaacson & Isaacson (1975, pp. 11–13) will substantiate the modification made here.
† Can also be expressed as mass rate of flow per unit area: $ML^{-2}T^{-1}$.
Sources: Huschke (1959), Lowman *et al.* (1972), Murphy (1949), and Strahler (1958).

numbers (Strahler 1980, p. 7). The underpinning of such an argument is disarmingly simple: the issues of interest to geomorphologists may be reduced to the dimensions of mass, length, time, and temperature that are found in classical Newtonian mechanics and thermodynamics (Strahler 1958, p. 281). In an updated review of these concepts, Strahler (1980, pp. 4–6) provided an extensive listing and a classification of such variables (Table 12.5). Once this proposition is accepted, the primary objective becomes to create geomorphic models in the form of rational equations. A

Table 12.6 Rational and empirical equations, plus a dimensionless number.* (From Strahler (1980, pp. 7–8).)

A rational equation
The following equation is an example:

$$Q = AV$$

where Q = discharge (having the units L^3T^{-1}), A = cross-sectional area of channel (having the units L^2), and V = mean velocity of flow (having the units LT^{-1}).
  Substituting dimensions the equation becomes

$$L^3T^{-1} = L^2 . LT^{-1}$$

When multiplied through the equation becomes

$$L^3T^{-1} = L^3T^{-1}$$

An empirical equation
The well-known Manning equation relates mean stream velocity (V) to hydraulic radius (R) and slope (S) as the expression:

$$V = 1/nR^{2/3}S^{1/2}$$

  If appropriate units are substituted in this equation the only way it can be balanced is to assign n (Manning's roughness coefficient) the units or dimensions of $L^{1/3}T^{-1}$. This is obviously not meaningful; consequently while it is an extremely useful equation, the Manning equation is empirical.

*A dimensionless number**
Strahler's (1958) geometry number was introduced to provide a fairly stable (one that hovered about unity) measure of landscape ruggedness. It is

$$HD/S$$

where H is local relief (units of L), D is drainage density (units of L^{-1}), and S is the tangent of ground surface slope (units of 1).
  A unit or dimensional interpretation of the equation is

$$\frac{L . L^{-1}}{1} = 1$$

---

* See footnote to Table 12.5.

rational equation is one in which the two sides of the equation are not only numerically correct, but also dimensionally correct. This means that when the dimensions on each side of the equation are summed (by addition, subtraction, multiplication, and division) they should balance. If they do not the equation is not rational, but empirical. Simple illustrations of rational and empirical equations, taken from Strahler, are given in Table 12.6.

  A situation in which the dimensions in an equation cancel each other out to produce a zero sum, or in which a dimensionless variable such as an angle is included (see Strahler (1958, p. 284) for an explanation), will produce a dimensionless number. The Froude and Reynolds numbers are well-known examples in fluvial geomorphology and are discussed by Strahler (1958); and the geometry number developed by Strahler (1958, p. 295, 1980, pp. 7–8) is illustrated in Table 12.6. One attribute of a dimensionless number is that, conceptually, it permits comparison of the geomorphic contexts without regard to scale differences. There are several dimensionless numbers among the 15 variables used by Melton.

A final and important point made by Strahler (1980, p. 11) is that it is inappropriate in his dimensionally balanced approach to integrate energy and material flow diagrams. This is because the two are not dimensionally compatible. When such integration is made, the result may present a satisfactory qualitative picture, but does not offer the opportunity for meaningful manipulation of data in accordance with the diagram outline.

Canonical structures

As may be seen in the case study a canonical structure is no more than a schematic diagram that represents a combined statistical (or indeed deterministic) and geomorphic interpretation of a system model. Such diagrams have been very widely used because they permit most of the salient points in a system model to be illustrated in a number of different ways simultaneously. They also afford the opportunity to emphasize linkage itself, again a vital component of a systems approach.

Melton's diagrams show most of the essential ingredients of a canonical structure. Small subsystems are gathered together, variable interrelations within the subsystem are emphasized by annotated lines indicating: (1) nature (direct or inverse); (2) direction (independent and dependent variable); (3) strength (one or more specified correlation coefficient confidence levels). By following the pathways of such linkages it is also possible to identify feedback relationships if they exist. Subsystems are demarcated by some sort of boundary marker, usually a box outline. This entire process is then repeated by linking and demarcating subsystems in the same fashion.

An important addition to Melton's diagrams is offered in those of Chorley & Kennedy, that is, identification of specific roles played by individual variables (Fig. 12.7). The most important new concepts are those of a regulator and a store. Chorley & Kennedy (1971, p. 354) defined a regulator as "a component that tends to stabilize the system internally", and a store as "a device capable of retaining mass, energy or information" (Chorley & Kennedy 1971, p. 356).

Chorley & Kennedy (1971, pp. 81–2) enumerated a number of regulators and stores. Many regulators exist as thresholds of one kind or another, e.g., slope angle or infiltration capacity. Some other regulators dispose of energy or mass by combinations of absorption, deflection, and/or reflection; for example, consider the complex impact of vegetation upon rainfall. Yet

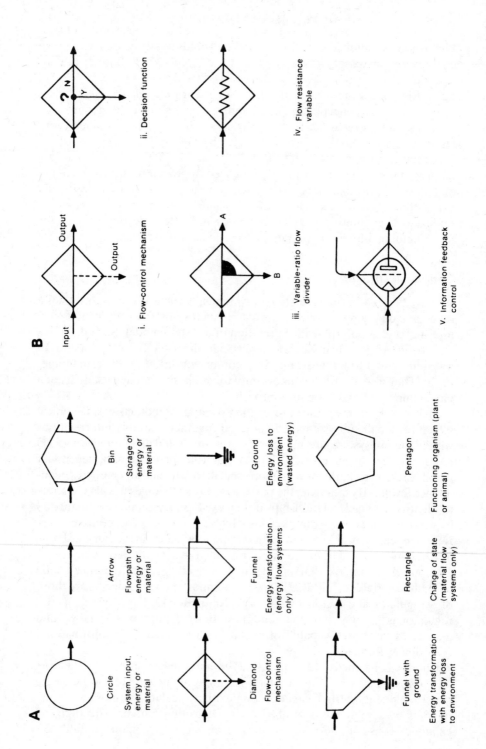

A

Circle — System input, energy or material

Arrow — Flowpath of energy or material

Bin — Storage of energy or material

Diamond — Flow-control mechanism

Funnel — Energy transformation (energy flow systems only)

Ground — Energy loss to environment (wasted energy)

Funnel with ground — Energy transformation with energy loss to environment

Rectangle — Change of state (material flow systems only)

Pentagon — Functioning organism (plant or animal)

B

Input   Output   Output
i. Flow-control mechanism

ii. Decision function

A   B
iii. Variable-ratio flow divider

iv. Flow resistance variable

v. Information feedback control

other regulators are significant because of their presence or absence, a basal stream at the foot of a slope (Fig. 12.6) being a classic geomorphic example. In a similar fashion stores may gather water (e.g., surface vegetation and ground water), energy (e.g., by a substance heating and cooling), and material of all sizes (e.g., coarse debris on a talus, fines in a lake sediment). The importance of a store is twofold because not only will it absorb (a positive function) but it will also release (a negative function).

Strahler (1980, p. 11) criticized Chorley & Kennedy's choice of symbols. His primary objection was to the limited yes/no capacity of their regulators. The yes/no dichotomy serves well in a human decision-making process, but does not permit the continuously variable regulation that frequently occurs in nature. Strahler preferred the symbols used by Odum in many papers focused upon ecological modeling. Perhaps the most useful reference of many is Odum (1972). Odum's symbols were derived from those used in electrical circuit diagrams; the basic elements as modified by Strahler (1980) are shown in Figure 12.8, and matched with one of Strahler's energy equations in Figure 12.9.

Canonical structures: geomorphic illustrations

Chorley & Kennedy (1971) present numerous examples of canonical structures featuring a wide range of geomorphic contexts. The double diagram in Figure 12.6 is one very interesting example taken from original work by Kennedy (1965). The same four subsets are considered in each diagram, namely those of debris, vegetation, slope geometry, and basal stream (i.e., the stream at the foot of the slope). The two diagrams contrast the relationships when the stream is actually flowing at the slope foot with those when the stream has moved away from the slope foot. Even a cursory comparison of the two diagrams creates a strong qualitative impression of

Figure 12.8 Symbols used by Strahler to depict functions in a canonical structure. Strahler's symbols are drawn largely from Odum (1972). (a) Basic circuit symbols for diagramming energy and material flow systems. (b) Flow-control mechanisms used in circuit diagrams of energy or material flow systems. These mechanisms divide the flow into two pathways (bifurcation) or impede flow in a single pathway. (i) General symbol for a *flow-control mechanism*. Input and outputs are at corners of the diamond. Within the diamond diagonals connect input to one or two outputs. (ii) *Decision function*, represented by a question mark. All flow is directed into one output or the other, depending on whether a query is answered as "Yes" or "No". (iii) *Variable-ratio flow divider* can divide the flow in any ratio between two outputs. A and B. Thus, output of A can vary from 0% to 100%, while output of B varies simultaneously from 100% to 0%. (iv) *Flow resistance* in a single pathway regulates the throughput. As resistance increases, a backup effect is propagated in a direction opposite to the flow direction. (v) *Information feedback control* is represented as a radio tube. A very small input of energy (as information) serves as a feedback to regulate the main flow of energy or material in a cybernetic system. (From Strahler (1980).)

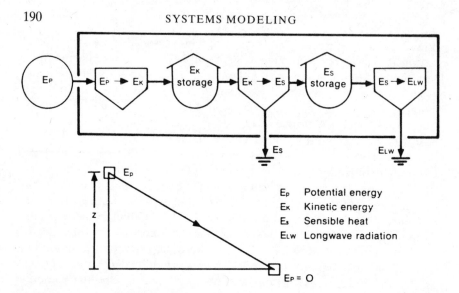

Figure 12.9 A stream segment represented as an open energy system model using symbols and terminology suggested by Strahler. The system model consists of an idealized reach of river channel of uniform gradient along which no discharge is added or lost. Energy loss by surface evaporation is assumed to be zero. At the upper end of the reach, potential energy, E_P, enters the system. Potential energy is equal to the product of mass (m), acceleration of gravity (g), and height (z) above the lower end of the reach. Rate of inflow of potential energy through the upper cross section of the stream is in units of power (energy per unit time, ML^2T^{-3}). Potential energy, E_P, is continuously transformed into kinetic energy, E_K, which enters temporary storage represented by the water in downstream motion, including turbulence. Kinetic energy, E_K, in turn is transformed into sensible heat, E_S, which continually escapes by conduction through the channel wall and to the air above. Sensible heat, E_S, is fed into storage, measured by water temperature, and is then transformed to longwave radiation, E_{LW}, which escapes into the atmosphere. Not shown is energy transformation into latent heat, E_{LH}, during surface evaporation. In steady state

$$dE_P/dt = dE_S/dt + dE_{LW}/dt$$

and energy in storage holds constant. At the same time, mean velocity and all velocity gradients are time-invariant, as are form elements such as cross-sectional area, width, depth, wetted perimeter, and form ratio. (From Strahler (1980).)

marked differences. It can be seen that this applies not only to relationships between subsets but also to relationships within subsets; the slope geometry subsets in the two diagrams exhibit some marked differences.

In overall terms, it is quite apparent that relationships are both stronger and more numerous where the stream is presently located at the slope toe. This may be interpreted as reflecting a system that is in some sort of functional equilibrium with processes actively operating upon and/or within it. Conversely, the lack of a web of strong correlations on slopes lacking a basal stream may be interpreted as indicating that such slopes lack a functional equilibrium, i.e., they are largely inactive and are adjusted to something that is not currently present, or they are in transition. It also follows that the length of relaxation time is a critical consideration. If such slopes attain equilibrium it is likely that this is done much more quickly when the stream begins to undercut the slope than when the slope begins to adjust to the absence of a stream. This reasoning stems simply from the energy available during the two periods, the normal argument being that a landform is adjusted directly to the energy exerted upon it, and inversely to the time since that process ended. Therefore, these two diagrams represent uncertain positions on a continuum.

Figure 12.9 summarizes Strahler's dimensionally correct equation for energy transfer through an idealized river channel segment. The figure is essentially self-explanatory. In the context of a rational or empirical equation the primary role of the symbols is to emphasize the role being played by the individual elements in the equation; the linkage and its direction being implicit in the mathematical manipulation.

Systems modeling: an evaluation

It is quite apparent that the primary strength of systems modeling is its capacity to direct the researcher's thinking toward the linkage, or organization, present in the topic under consideration. The fact that this perspective, once developed, may be sustained through increasingly sophisticated techniques (e.g., soft system, correlation approach, empirical equation, rational equation) lends a consistency and a flexibility that are also very appealing. In brief, a holistic approach with sufficient flexibility to absorb many different techniques is the most appealing virtue of systems modeling.

Another virtue of system models is that they utilize symbols. This is a form of simplification, but a powerful one because it represents abstraction. In general, when such simplification and abstraction are present, mathematics can be applied to the issues at hand. Therefore, a general suitability for various types of mathematical expression may be cited as a specific example of the more general point made in the preceding paragraph.

The primary weakness of a systems approach is that the researcher manipulates only what he or she chooses to include initially. Obviously, any answer provided by a system model can only be derived from the inputs; therefore, both inclusions and omissions are critical – but there is no formal procedure for generating this selection. Accepting the notion that the ability to falsify is the keystone to science, this initial step in systems modeling is unscientific.

The same problem may be expressed, not in terms of variables, but in terms of boundaries. It is very apparent that most systems modeling in geomorphology focuses upon individual landforms and attempts to explain them in terms of form-process interaction. This means that the core of the issue is usually very clear, but the marginal issues (be they expressed in terms of which variables to include and exclude, or in terms of where to place system and subsystem boundaries) are extremely tenuous and potentially critical. The very nature of models, the fact that they are simplifications, means that this particular problem is unlikely to be eliminated.

Most system models as presented in geomorphology emerge as summaries, or statements of the manner in which things are related. Conversely, the structures of these models are rarely related strongly to their theoretical origins. Little or nothing is learned by relating things in a superficial manner and/or by calling things already known a system. The real strength of a systems approach should be that it reveals unappreciated relationships and that the model's structure is then related to theory by specific tests. In brief, system models too frequently emerge as conclusions or statements of fact, when they should be presented as testable hypotheses. This is not so much a negative criticism of systems modeling *per se*, but rather of the way in which geomorphologists have used it.

The final point is merely a reminder, but seemingly a very necessary one. System models are models; they are intellectual and simplified constructions – not reality. Therefore, they must be tied very closely to existing theory and their frailties kept very much in mind. A specific illustration of the need for careful model construction appeared in Murdoch's (1966) appraisal of trophic levels. A more broadly based, but equally insightful, critique of the dangers of being swept uncritically into the tide of systems modeling in geomorphology, and environmental sciences at large, may be found in Kennedy (1980). In short, the fundamental case for systems modeling rests on a holistic perspective; that against the technique stems from a reductionist perspective.

13 Mathematical models*

Despite the common use of the word "theoretical" to describe what is actually "mathematical" geomorphology, there is nothing inherently theoretical about mathematics. In fact mathematics is simply a language, and as such may be viewed as no more, nor less, than an alternative to the written word. All languages differ in their strengths and weaknesses; the great strength of mathematics is that it permits precision, thereby eliminating ambiguity. A fundamental weakness of mathematics is the precision with which mathematical arguments must be established, because, implicitly, this requires great knowledge of the topic to be treated. An extension of this issue is the need for highly refined data as inputs to most numerical computations. Finally, the very same lack of flexibility cited as a strength may become a weakness under some circumstances.

As a result of this pattern of strengths and weaknesses, mathematical models usually exhibit internal logic, i.e., once begun each step leads logically to the next. It is true that such models may contain errors, but this is not generally their primary source of weakness. The most likely source of error in mathematical models is in their external logic, i.e., in the assumptions that were made to permit the model to be created in mathematical form. Specification of a geomorphic situation with sufficient precision to make it mathematically tractable often requires very constraining assumptions, if not some wholly unrealistic ones.

In choosing to use the language of mathematics the most important decision forced upon the researcher is whether to view geomorphological systems and/or processes as deterministic or probabilistic (Kirkby 1976a, p. 3). In the first option basic (e.g., chemical, physical) geomorphic controls are viewed as being a set of unique responses; therefore, if condition 1 exists, condition 2 and no other will follow. In the second option the notion of randomness is invoked, so that an element of variability is introduced. A third, seemingly intermediate position, is actually tenable. In this, the researcher accepts that the world is actually deterministic, but believes it to

* The research and mathematical appraisals for this chapter were undertaken by Deborah S. Loewenherz (University of Illinois). Her work, which made this chapter possible, is gratefully acknowledged. However, any errors or omissions remain my responsibility.
C.E.T.

be so inordinately complex that the only feasible approach is to treat it stochastically. In fact, this third position is very appealing at present because it allows the researcher "to keep the faith" (i.e., maintain a deterministic intellectual posture) while using a somewhat less demanding research approach as befits a limited grasp of many complex issues.

Geomorphologists with mathematical training are at a distinct advantage over those who do not have one. This primarily stems from the access such training gives them to the main body of scientific literature, and the opportunity it offers to conduct their own research in a similar, or at least compatible, fashion. However, the majority of geomorphologists still lack an adequate mathematical background. The net result is divergence; in an already small community, two groups (the mathematical and nonmathematical) are developing using different languages. The mathematical group is able to understand the nonmathematical one, but the inverse is not true. This situation is exacerbated by the mystique with which the nonmathematicians endow their mathematical colleagues. The intellectual issues are not necessarily enhanced by the choice of language with which they are addressed, nor are the answers necessarily more refined. Nevertheless, despite the present feasibility of functioning as a nonmathematical geomorphologist, the desirability and potential longevity of such an academic lifestyle in research geomorphology must be seriously questioned. Indeed, any student would be foolish to embark upon a course of training in geomorphology that did not include a sound foundation in calculus and statistics.

Most mathematical treatments of slope evolution identify the pioneering paper of Lehmann (1933) as the starting point of modern mathematical efforts. This theme was picked up and taken further considerably later in a series of papers by Bakker and LeHeux (1946, 1947, 1950, 1952). Another concerted effort to develop a comprehensive mathematical model was that of Culling (1960, 1963, 1965); while Scheidegger (1961), who is perhaps best-known for his later textbook (Scheidegger 1970) on the topic, was also an early worker in the field. These initial models were often quite simple in their mathematical approach, but it is important to appreciate the difficulty of developing such ideas when they were so much out of step with the mainstream of geomorphology. This chapter is an attempt to illustrate mathematical geomorphology by examining and comparing the models of two geomorphologists who have developed and refined their models over the course of many years, Frank Ahnert and Michael J. Kirkby.

Frank Ahnert

Scientific context

Although initially trained in his native Germany, Frank Ahnert served on the faculty of the University of Maryland for many years. This is significant because his approach to geomorphology is much more in tune with the North American/British school of process geomorphology than it is with the German school of climatic and genetic geomorphology. So while Ahnert's use of mathematical models is still a minority choice even in English-language geomorphology, his intellectual roots may be seen as resting firmly within contemporary English-language geomorphology.

A mathematical equilibrium model

Ahnert (1967) introduced his modeling approach in a combination of nonmathematical text and simple, conceptual equations. This provides an excellent starting point that should be within everyone's grasp. The central concept is that of equilibrium, traced back to Gilbert (1877) but, unlike Hack (1960), interpreted in terms of the balance between waste supply and removal. Ahnert (1967, p. 25) specified some important constraints and expectations of his model: (1) it must be practically applicable (i.e., potential field verification must be feasible); (2) it must be comprehensive (i.e., contain slopes and streams, as well as erosional and depositional processes and forms, and demonstrate landform change through time); (3) initial assumptions must be minimal; (4) it must be generally applicable, but also applicable to specific cases; (5) it should be geographical (i.e., explain spatial variation in processes and landforms). In short, it must be a genuine attempt to produce a model of worldwide validity.

Ahnert started from an elementary, but eminently defensible, proposition (Ahnert 1967, p. 25):

> One may start with the simple recognition that the configuration of landforms results from the different rates at which rock waste is produced, transported, and redeposited at different points of the surface, i.e. from the spatially (and temporally) varying rates of waste supply and removal.

From this general statement Ahnert immediately suggested three ways in which the rate of waste production from bedrock may be related to the depth of overlying waste (Table 13.1). He thought that Equation 1 might be a realistic one for modeling mechanical (freeze–thaw) weathering because temperature oscillations across 0°C decrease with depth; and that Equation 2 might be satisfactory for some, but not all, chemical weathering processes, but that it is flawed because it implies that bare bedrock is an optimal condition for chemical weathering while, in truth, a thin waste cover is optimal. Equation 3 is an illustration that real-world complexity can be modeled mathematically, as it incorporates the weathering patterns of both preceding equations. The waste cover at any point on a slope is then considered in terms of production, delivery, and removal and a number of

Table 13.1 Possible relationships between rate of bedrock waste production and depth of overlying waste mantle. (After Ahnert (1967, p. 26).)

Terms

C	=	waste cover thickness
W_C	=	rate of bedrock wasting under a waste cover thickness of C
W_0	=	rate of bedrock wasting on a bare rock surface
D_1	=	limiting waste cover thickness beyond which the negative linear function (see Equation 2 below) retains the value of 0
e	=	base of natural logarithms
p	=	proportionality factor $(0 \leqslant p \leqslant 1.0)$, which will vary in magnitude with climatic and petrographic conditions

Equations

Equation 1 describes an inverse exponential relationship between the bedrock waste production rate and the thickness of the overlying waste:

$$W_C = W_0 e^{-C} \tag{1}$$

Equation 2 describes a negative linear function, and is only valid for $C \leqslant D_1$:

$$W_C = W_0(1 - C/D_1) \tag{2}$$

Equation 3 describes a combination of Equations 1 and 2:

$$W_C = W_0[pe^{-C} + (1-p)(1-C/D_1)] \tag{3}$$

specific conditions identified and defined (Table 13.2). As weathering (W) is a function of waste cover thickness (C), if the waste cover thins weathering increases; while if waste cover thickens weathering decreases. This has been illustrated diagrammatically by Carson & Kirkby (1972, p. 105) (Fig 13.1). As Ahnert pointed out, his framework exhibits true dynamic equilibrium, with any change leading to movement towards re-establishment of equilibrium, while also permitting such trends to exhibit feedback relationships.

As presented so far, this is a model of process at a point, but it must be converted to an expression of form. This is done by applying exactly the same set of ideas to an entire slope (Fig. 13.2), including some special cases (Fig. 13.3). The combination of what might be called conceptual equations and diagrams clearly provides a basis for quantitative modeling, providing that the various parameters may actually be approximated on the basis of field observations. Of the specified parameters, the arrival and removal of surficial waste are by far the most readily measured, while actual changes in form occur so slowly that their measurement may be almost impossible on most slopes. However, Ahnert (1967, p. 38) noted that these rates may be related to existing form and this would then permit substitution. This suggestion is founded on an equilibrium concept and proposes an ergodic solution; however, given a quantitative approach, both notions might eventually be tested formally. Ahnert (1967) also treated stream profile evolution fully, but this aspect of the paper will not be pursued here.

Table 13.2 Conceptual outline of possible relationships between waste production, delivery, and removal at a point on a slope. (After Ahnert (1967, pp. 26–8).)

Terms

C	=	waste cover thickness at the beginning of a unit time period
C'	=	waste cover thickness at the end of a unit time period
A	=	rate at which waste arrives from upslope
W	=	rate of waste production by local weathering
R_p	=	potential rate of waste removal

Development
General relationship

$$C' = C + A + W - R_p$$

Equilibrium

$$R_p = A + W \qquad (\text{i.e. } C' = C)$$

Equilibrium subtypes:
"denudational equilibrium" occurs if $W > 0$ and $R_p = A + W$ (i.e. the slope is being worn down, but its waste cover thickness remains unchanged)
"transport equilibrium" occurs if $W = 0$, then $R_p = A$ (the waste cover is so thick that weathering has ceased and the slope point is not being lowered)
Disequilibrium

$$C' \neq C$$

Disequilibrium subtypes:
"negative disequilibrium" occurs when $C' < C$, that is, $R_p > A + W$
"positive disequilibrium" occurs when $C' > C$, that is, $R_p < A + W$

Note: Traditional "degradation" occurs whenever $R_p > A$; it must occur during denudational equilibrium and negative disequilibrium, but may also occur during positive equilibrium.
Traditional "aggradation" can only occur during positive disequilibrium when $R_p < A$.

The full mathematical development of Ahnert's slope model appeared in a series of papers (Ahnert 1971, 1973, 1976, 1977). In the first paper Ahnert (1971) examined slope profiles. This kind of model is usually called two-dimensional by geomorphologists, but one-dimensional by the engineering and geotechnical community. The latter designation is based on the notion that there is only one independent spatial variable in a slope profile model. In other words, variations in height are examined in one direction only, along a fixed x-axis. In the engineering and geotechnical literature, a two-dimensional slope model would represent variations in height throughout a fixed xy surface. Ahnert's model is discrete, which means that it comprises a series of points and computations for each point are made individually and sequentially. In order to make a discrete model as accurate as possible it is necessary to use small spatial and temporal increments. As the model is founded on calculation of changes in waste cover thickness (or material balance to express it another way), Ahnert had to develop

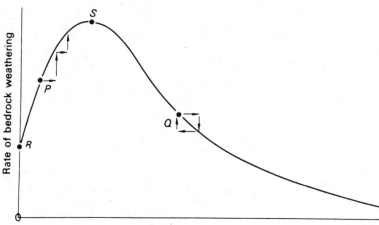

Figure 13.1 Schematic illustration of bedrock weathering rate as a function of soil depth. The optimal rate occurs at S. To the left of S the soil is thinner than at the optimal rate; therefore, if soil thickness increases so does bedrock weathering rate – a positive feedback loop or instability prevails between R and S. To the right of S bedrock weathering rate is below the optimal rate because the soil thickness is too great; therefore, anything that tends to increase soil thickness tends to reduce the weathering rate – a negative feedback loop or stability prevails. On the left-hand side of R slopes are largely weathering-limited; to the right of R they are mainly transport-limited; and the two types exhibit little overlap. (From Carson & Kirkby (1972).)

equations to describe all of the weathering and transport processes he wished to include. A general flow diagram for the entire model appears in Figure 13.4.

Ahnert used three equations for weathering: (1) a mechanical weathering function which decreases exponentially with depth (i.e., the original Equation 1 in Table 13.1); (2) a chemical weathering function that permits chemical weathering to increase initially as waste cover thickens, but then makes it decrease after a specified thickness has been exceeded; (3) a combination function which melds (1) and (2). The last point is interesting because the proportions of (1) and (2) which are set in (3) are chosen by the researcher; therefore, they are external to the model itself but permit climatic influences to be modeled.

Ahnert (1973) is largely a computational improvement over the earlier (Ahnert 1971) paper; the Fortran program was cleaned up and made more efficient while the ideas remained largely unchanged. The 1976 version of the model (Ahnert 1976) contains expansion of the model to three dimensions (two dimensions in engineering and geotechnical terminology). This expansion was fairly easy mathematically because Ahnert's model was not based on a continuous differential equation; therefore, replacement of a

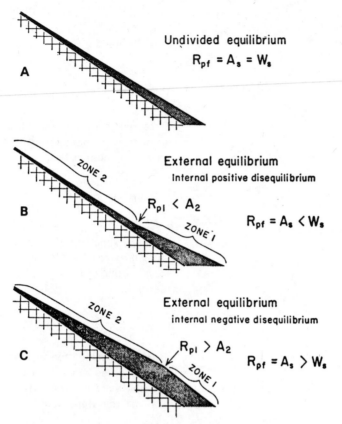

A Undivided equilibrium
$$R_{pf} = A_s = W_s$$

B External equilibrium
Internal positive disequilibrium
$$R_{pl} < A_2$$
$$R_{pf} = A_s < W_s$$
ZONE 2
ZONE 1

C External equilibrium
internal negative disequilibrium
$$R_{pl} > A_2$$
$$R_{pf} = A_s > W_s$$
ZONE 2
ZONE 1

Figure 13.2 Possible equilibrium conditions over an entire slope.
(a) *Undivided equilibrium* occurs if the potential rate of waste removal at the slope foot (R_{pf}) equals the rate of waste arrival at the slope foot (A_s), and if in turn A_s equals the rate of waste production on the slope (W_s).

(b) *External equilibrium (internal positive disequilibrium)* occurs when some environmental change produces a decrease in R_{pf} and waste accumulates at the slope foot. Initially $R_{pf} < A_s$ and the slope angle decreases; in turn, this will result in a decrease in A_s until equilibrium ($R_{pf} = A_s$) is restored at the slope foot. However, while the upper portion of the slope continues to experience the old weathering rate $A_s < W_s$ the slope is in external equilibrium, but internal disequilibrium, and divided into a lower zone (zone 1) adjusted to a slower rate of waste removal at the slope foot, and an upper zone (zone 2) that has yet to respond to the change. At the boundary between the two zones $R_{pl} < A_2$ and aggradation occurs. The boundary exhibits a concave knick in the profile, which moves upslope until zone 1 covers the entire slope. Bedrock weathering at the slope foot slows as waste accumulation thickens, but continues at the old rate on the upper slope; the net result is a more convex bedrock slope profile.

(c) *External equilibrium (internal negative disequilibrium)* occurs when $R_{pf} > A_s$. It produces thinning of the waste cover and an increase in weathering rate at the slope foot until the external equilibrium is regained. As long as $A_s > W_s$, internal negative disequilibrium is maintained. This may be expressed as $R_{pl} = A_1 > R_{p2} = A_2$ with $R_{pl} > A_2$ occurring at the boundary between the zones. Again the boundary shifts upslope until undivided equilibrium is restored. (From Ahnert (1967).)

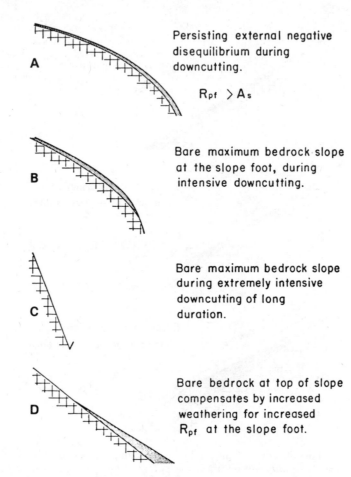

A Persisting external negative disequilibrium during downcutting.

$$R_{pf} > A_s$$

B Bare maximum bedrock slope at the slope foot, during intensive downcutting.

C Bare maximum bedrock slope during extremely intensive downcutting of long duration.

D Bare bedrock at top of slope compensates by increased weathering for increased R_{pf} at the slope foot.

Figure 13.3 Some special slope conditions.

(a) Persisting external negative disequilibrium during downcutting. As a river cuts down at the slope foot it maintains a waste removal rate (R_{pf}) that is greater than the supply rate (A_s), even though A_s is increasing. The result is development of a series of convexities at the slope foot, each of which advances upslope; the net result is an overall convex profile.

(b) Bare maximum bedrock slope at the slope foot, during intensive downcutting. Intensive downcutting by a stream exposes bare rock at the foot of the slope. The bare rock will maintain the maximum angle it can hold and there will be a hanging waste cover higher upslope.

(c) Bare maximum bedrock slope during extremely intensive downcutting of long duration. Very intensive downcutting is maintained over a long period of time. This causes the bare bedrock, which is maintaining its maximum angle, to extend all the way to the slope crest.

(d) Bare bedrock at top of slope compensates by increased weathering for increased R_{pf} at the slope foot. If the rate of bedrock weathering is unable to compensate for an increase in the rate of removal at the slope foot, then bare bedrock first becomes exposed at the slope crest. Bedrock exposure then advances downslope until its increased waste production fully compensates for the increased rate of removal. Such a bare bedrock slope may not be at the maximum angle possible, although it may be cliffed. (From Ahnert (1967).)

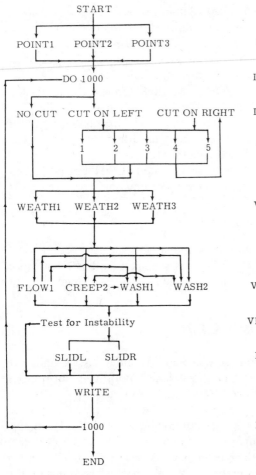

I. Start of Program.

II. Read-in of the initial profile.

III. Begin of main loop.

IV. Decision whether (and where) to cut down.

V. Modes of Downcutting: 1. uniform; 2. linearly waxing; 3. geometrically waxing; 4. lin. waning; 5. geom. waning.

VI. Modes of weathering (and geol. structures): mech., chem., and combined.

VII. Modes of waste transport (used singly or in combination)

VIII. If local slope is steeper than 45 degrees, waste will slide.

IX. Landslide mechanism

X. Write-out of parameters.

XI. End of main loop.

Figure 13.4 A generalized flow chart for Ahnert's COSLOP model computer program. (From Ahnert (1971).)

series of points forming a profile by a grid of points forming a three-dimensional form merely means that more points must be calculated, but still individually and sequentially. As there are several important modifications from the 1971 model version, it is the transport equations of the 1976 version that will be considered here.

One of the most important attributes of Ahnert's model is that he distinguished between point-to-point transfer and direct removal. In the former, material removed from one point is delivered to the next point downslope, but in the latter material removed from a point is lost to the system without being redeposited. Point-to-point removal is the norm and it always tends to produce accumulation at the base of a slope. Ahnert also specified slow mass movement as viscous (nonthreshold) or plastic (threshold) processes. Standard equations for nonthreshold processes specify that the transport capacity of the process will increase as slope gradient increases and/or as distance from the divide (slope crest) increases. Equations for threshold processes are formulated in much the same fashion as those for nonthreshold processes, except that a critical slope gradient value must be exceeded before the slope process is initiated. Mathematically, the nonthreshold equation is merely a special case of the threshold equation in which the critical slope angle is zero. Ahnert used five transport process equations. A brief comment on each equation follows.

Splash transport was modeled on ideas presented by DePloey (1972) and is made solely dependent on slope gradient. The equation is

$$R = K_2 \sin^m \alpha$$

where m and K_2 are constants and α is slope angle.

Wash transport (Ahnert 1971, 1977) was specified as dependent upon slope gradient, surface runoff, and erodibility of surface material. The equation is

$$R = (K_3 + K_4 C) D^n \sin^p \alpha$$

where D is surface runoff at a point, C is erodibility, and K_3, K_4, n, and p are all constants. Wash transport may be specified as a point-to-point transfer or as a process of complete removal (see above for discussion).

A viscous flow function was used by Ahnert to characterize slow mass movement (e.g., soil creep) which does not involve a threshold. His model (Ahnert 1971) contained a simple proportionality between transport rate and slope gradient. However, the later papers (Ahnert 1976, 1977) contain a modification based on suggestions by M. A. Carson which accounts for the "fluidity" of the material. The equation is

$$R = K_5 C^r \sin \alpha$$

where K_5 and r are constants representing "fluidity". Ahnert (1976) introduced plastic downslope flow as a type of mass movement. Computationally it does not differ very much from viscous flow, except that a critical slope gradient value must be exceeded (i.e., it is a threshold process).

In order to simulate landscape development realistically, Ahnert (1971, 1976, 1977) constrained his numerical model so that slope gradients above a specified value do not maintain a waste cover. Once the specified value is exceeded at a point, material is moved downslope until it arrives at a point where the constraining condition is met. This satisfies the need of the model, but it is not a realistic description of landsliding. The fundamental problem is that landsliding is not a slow mass movement process but a rapid one; therefore, travel distance may actually be much greater than the model computes. In more fundamental terms, the critical threshold associated with initiation of movement may vary from point to point within the landscape as potential energy is converted to kinetic energy. The net result is that the momentum of movement may produce a significant effect itself.

Ahnert's (1967) initial model contained no explicit slope hydrology component, although it is clearly necessary for the calculation of surface wash. His later versions (Ahnert 1971, 1976, 1977) incorporated a balance equation to determine runoff values at each point. The equation utilizes known precipitation values at particular points, plus infiltration values and runoff from upslope. The model is also flexible enough to include variation in rock structure and base level. The later model versions simulate lithology by permitting varying resistances to be assigned to any point, while base level changes are simulated by changing the reference datum between iterations (computer runs).

The approach that Ahnert has developed is heavily oriented toward numerical simulations. This means that specific values may be entered and the researcher may watch the slope profile, or landscape segment, change through time in accordance with the values inserted. In the 1971 and 1976 model versions the researcher may specify the initial surface configuration, proportion of mechanical to chemical weathering, nature of base level changes, and the type of transport dominating mechanical transport. Such manipulations were illustrated by Ahnert (1971, 1976) and the results led him to conclude that the main qualitative differences in slope form are derived not from the specific transport type, such as plastic versus viscous movement, but from the mode of removal (i.e., direct removal versus point-to-point transfer). In a recent paper, Ahnert (1987) has published some revisions of the 1976 version of the model and used the updated model to illustrate both the general nature of dynamic equilibrium and slope development in the Karl valley, West Germany.

Ahnert's model exhibits some obvious strengths, including: his identification of waste cover as an important factor in landscape development; his use of waste cover as the linchpin that permitted him to verify empirically the validity of his model; his later model versions also permitted evaluation of many process–form relationships. The two primary weaknesses in the model are that: (1) it is not physically based in all instances (i.e., the equations are not directly derived from established chemical and physical principles); (2) the various process functions in the model are not coupled – in other words they operate independently and there is no feedback effect between the various processes. These limitations should not be allowed to mask the fact that this model is a tremendous pioneering effort.

Michael J. Kirkby

Scientific background

Kirkby is one of several leading geomorphologists to have been trained in the Department of Geography at Cambridge University during the period that R. J. Chorley has been the most influential geomorphologist there. Chorley has pursued a systems approach much more consistently than any other; however, Kirkby's early mathematical training has remained evident throughout his career and he has developed deterministic modeling to perhaps a higher level than anyone else. Thus, his background places him in the center of the process–response approach to geomorphology, but his own work appears to have been stimulated as much by pioneering papers by Culling and others as it does from any other source.

A deterministic continuous slope model

Kirkby (1976a, pp. 2–3) has discussed the possible levels at which slope models may be conceived and executed. He provided a skeletal diagram (Fig. 13.5) to illustrate how process and form interactions are related to other fundamental geomorphic issues. Comprehensive modeling of this set of relationships forms the goal of Kirkby's model, although it was not all achieved at one attempt. Kirkby (1971) introduced a model founded on a differential equation that constrained mass balance (or material balance), and thereby maintained continuity and simultaneity of the slope. This mathematical approach means that the entire slope profile is modeled, unlike Ahnert's model where only a series of points on the profile were calculated. Kirkby's initial model was constructed so that submodels could be absorbed as they were created. Kirkby also presented the model using vector operators so that the equation is independent of any particular coordinate system. Vector notation allows a governing equation to be expressed in a general form, which may be used to describe a one-, two-, or three-dimensional system. In order to solve such an equation a coordinate system must be selected and the equation must be expressed within this particular spatial framework.

In his early work Kirkby relied exclusively on analytical techniques, but in recent work he has used numerical techniques to solve otherwise intractable equations. In an analytical approach the mathematician makes assumptions based on theory and expresses them symbolically (often algebraically). New equations are then derived by manipulating the symbols; i.e., it is very much a deductive methodology. However, when considering complex issues, theory often becomes inadequate for full development of interests at hand. The modeler may then turn to empirical (derived from experiment) data. If this is done, it becomes essential to insert values (numbers) into the equation, because there is no known

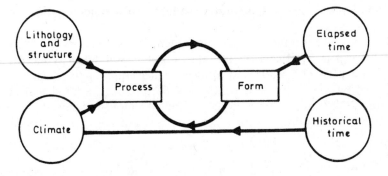

Figure 13.5 Relationships between slope models (process and form in the diagram) and independent variables. (From Kirkby (1976a).)

theoretical value that may be derived and assigned a symbol – only measured values which are nontheoretical estimates. Once an equation contains such a numerical value (often seen as a "constant" or "factor") the only answers that can be derived from the equation emerge as numerical values. Such an approach is a numerical technique and, while it may provide feasible results, it is not wholly founded on theory.

Kirkby's (1971) basic equation constrained sediment movement through the slope system, describing it continuously and in vector form. It is

$$\nabla \cdot (\mathbf{S} + \Delta) + dz/dt = 0$$

where (Kirkby 1976a, p. 10) $\mathbf{S}$ is a vector representing mechanical transport and Δ is a vector representing chemical transport. ∇ is called a "del operator"; when h is a vector $\nabla \cdot \mathbf{h}$ is the divergence of the vector $\mathbf{h}$; when h is a scalar (i.e., a quantity having magnitude, but no direction) then ∇h is the gradient of the scalar h. So $\nabla \cdot (\mathbf{S} + \Delta)$ is the divergence of the vector quantity formed by adding the chemical and mechanical transport vectors. dz/dt represents the change in elevation over time. Unlike Ahnert's equation, Kirkby's merely describes elevation change and does not distinguish the waste cover itself. The equation requires process functions to be developed which will describe weathering and transport. The explanations in the original paper (Kirkby 1971) are not particularly clear and much better ones are to be found in Carson & Kirkby (1972, Appendix B, pp. 433–6).

Kirkby (1971) considered weathering in terms of the relationship between mechanical removal and total removal (mechanical plus chemical) at a point on the slope. He used this concept to evaluate "equilibrium soils" in which he considered the degree of development to be proportional to the relative magnitudes of mechanical and chemical lowering (Kirkby 1971, p. 18). As noted by Kirkby, the relative rates of mechanical and chemical lowering differ down a slope profile, so that a soil catena is implicit to this model.

Modeling of weathering has subsequently been greatly extended by Kirkby: in Kirkby (1976a) by introduction of a "soil deficit" concept, and again in Kirkby (1977, 1985) by using a balance equation to account for the material budget of the soil. Kirkby (1976a, pp. 14–15) noted that everywhere within a soil some material has been removed chemically in comparison to the bedrock from which it is derived. He then described soil deficit (Kirkby 1976a, p. 15):

> This soil deficit may be thought of as the amount of material required to reconvert the soil into bedrock. It clearly increases with both soil depth and with degree of weathering, and seems to be the only unambiguous measure of amount of soil in a continuously varying soil profile.

Soil deficit may be expressed in a number of different equation forms, depending on the assumptions invoked, and Kirkby (1976a, p. 15) illustrated these, including a continuity form. In his later paper (Kirkby 1977) he offered the following balance equation for soil deficit in a simple soil profile:

$$dw/dt = d/dx[J - S(1-P_s)/P_s]$$

where x = distance from the divide, t = time elapsed, S = mechanical sediment transport measured in the x direction, J = chemical sediment transport, w = the total accumulated soil deficit, and P_s = properties remaining in the zone of mechanical erosion (basically this is the proportion of bedrock remaining). The soil profile model in this context is primarily chemical in nature and is strongly dependent on the effects of hillslope hydrology on chemical weathering and solute transport.

In two papers Kirkby (1976a, b) specified that chemical removal is proportional to the volume of subsurface flow. However, in Kirkby (1985, p. 215) he considered three possible functional relationships between subsurface flow and chemical removal. He selected a form that is a modification of his 1976 model function. The old relationship specified a constant proportion between subsurface flow and chemical removal, but in Kirkby (1985) he specified that the rate of change in chemical removal be constant with respect to changes in the volume of subsurface flow. Under the old specification it was possible for physically impossible discontinuities to emerge, a situation that is eliminated under the new specifications.

Kirkby (1971, p. 18) credited Gilbert (1877) with introducing the concepts of transport- and weathering-limited removal. In the case of slopes whose development is transport-limited, the actual transport rate (S) is equal to the transporting capacity of the processes (C). This means that, if the potential weathering rate is greater than the capacity of the transport processes, soil will deepen until it is so deep that the weathering rate is reduced below the potential rate. Conversely, if the potential transport rates are much greater than the rate at which soil is being produced by weathering, the slope will be bare bedrock and its development will be constrained by the weathering rate; it is a weathering-limited slope. Kirkby (1971, p. 19) added an important third category which he labeled "erosion-limited removal". He defined this state as follows:

There is perhaps an intermediate condition, applicable to cases where the depth of operation of the transporting process is very variable, for example river bed-load transport. In this case there is some surplus of transporting capacity over volume of unconsolidated material available for removal, and the erosion rate, $-dy/dt$, is assumed proportional to this surplus.

This concept may then be expressed as erosion-limited removal:

$$-dy/dt = k(C - S)$$

where everything is defined as above, and k is a constant of erosion. When k tends toward infinity, then $C = S$, and there is a transport-limited situation; when k tends toward zero, then $C \gg S$, and a weathering-limited situation emerges. Thus, the big advantage in introduction of the erosion-limited concept is that it permits all three situations to be encompassed by the same equation. Kirkby (1971, pp. 19–20) also incorporated threshold and non-threshold concepts into his model; mathematical formulation for the two is identical, because the latter is simply a special case of the former where the critical angle is zero.

Kirkby and Ahnert treat splash transport in essentially the same fashion. Kirkby (1976a) treated splash as a special case of mechanical transport; the general form of the equation is

$$S = F(q_{OF}, dz/dx)$$

where F means "a function of", q_{OF} is overland flow, and dz/dx is a slope gradient. This equation implies that the mechanisms are exclusively dependent on slope gradient, and in the case of rain splash becomes even simpler because q_{OF} is omitted. This seems to include a simplification because overland flow will presumably provide some protection against splash effects.

Initially Kirkby (1971) modeled wash transport as a simple, nonthreshold process that depended upon both distance from the divide (a surrogate measure of surface runoff) and slope gradient. In his later work (Kirkby 1976a, b, 1985) he created and incorporated a hydrological submodel. In the submodel, wash transport was respecified as proportional to slope gradient and the volume of overland flow, producing the equation (Kirkby 1985, p. 223):

$$S = q^m i^n$$

where i represents slope gradient, q represents overland flow, and m and n are both constants.

Kirkby (1971) originally treated slow movements with a standard, nonthreshold equation. Transport rates are solely dependent upon slope gradient, and distance from the divide alone does not influence the rate of movement or amount of material transported. In more recent work (Kirkby 1985) the function is constrained so that the depth of influence of the slow mass movement is taken into account.

Kirkby (1971, pp. 19–20) presented a general threshold transport law which incorporates viscous flow as a special case (i.e., where the threshold angle is set to zero). This equation is

$$C = \begin{cases} f(a)(-dy/dx - \tan\alpha)^n & \text{for } -dy/dx \geq \tan\alpha \\ 0 & \text{for } -dy/dx \leq \tan\alpha \end{cases}$$

where $-dy/dt$ is the erosion rate, C is transport capacity, a is the unit area drained per unit contour length, $f(a)$ is a positive function of a describing the influence of increasing area or distance from the divide, n is a constant exponent describing the influence of increasing gradient, and is usually set to zero or positive, and α is a constant angle ($0 < \alpha < 90°$).

In his early paper Kirkby (1971, p. 20) treated landsliding as a threshold-based, erosion-limited process. Subsequently, Kirkby (1985, pp. 225–8) re-examined the topic at considerable length and developed an expression for the distance traveled by a landslide based on the slope gradient. In fact, the derivation used does not directly model the physical processes acting, but it does take into account the potential energy characterizing the landslide and its effect upon the denudation rate.

The 1971 version of Kirkby's model contained no explicit hydrology component, but in two later papers (Kirkby 1976a, b) he developed a hydrology submodel. The submodel uses known values of annual precipitation, mean rain per day, and surface storage capacity. Using these values Kirkby was able to calculate a hydrologic balance which accounted for losses through evapotranspiration and distributed remaining moisture between overland and subsurface flow. By hydrological modeling standards the model is fairly simple, but it represents a significant effort in terms of slope modeling as illustrated in Figure 13.6. One important, implicit step forward with inclusion of a hydrological submodel is that it incorporated climate into the overall model (see Figure 13.5 for the significance of this addition).

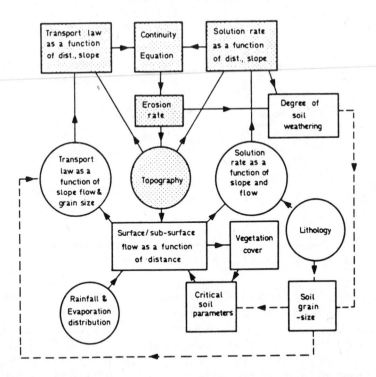

Figure 13.6   A schematic diagram of the impact of adding a hydrological submodel to slope modeling. As Kirkby pointed out, up to the time of the model under discussion, attempts at slope models had been in terms of topographic variables (mainly slope gradient and distance from divide) – these are shown in the stippled boxes in the diagram. Addition of a hydrologic submodel (aimed at incorporating climate into slope models) means that the model was extended, in principle, to include all of the topics within the boxes. (From Kirkby (1976b).)

    An ability to handle the effects of base level changes was introduced to Kirkby's (1976b) model when he respecified the mass equation. The continuity equation was no longer set equal to zero, but equal to a function of an independent variable in the equation (in this case, time). As a result, continuity is maintained subject to the rate of base level change as an additional constraint. However, in order to obtain an analytical solution more readily, Kirkby (1976b) assumed that base level remained constant (i.e., it occurs as a zero value in the equation). Much more recently, Kirkby (1985) has added lithological variability to his model. This has forced him to rely on numerical, rather than analytical, solutions to equations. Such a step seems intuitively reasonable because lithology is so complex that it is likely to resist a comprehensive understanding in purely theoretical terms.

    In its initial form Kirkby's model was a slope model; he called this a one-dimensional model as material moves in the x direction and produces

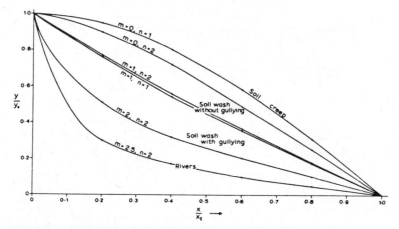

Figure 13.7 Dimensionless graph showing approximate characteristic-form slope profiles for a range of processes. (From Kirkby (1971).)

change in elevation (z) at a point. As noted earlier, the original notation is in vector form and may be readily expanded to two dimensions. The primary restriction on such a development was Kirkby's concentration upon analytical techniques. This approach was also founded on the belief that two-dimensional models would only become meaningful when the question of drainage initiation could be addressed (Kirkby 1976a, 1986).

Kirkby (1986) addressed drainage initiation by generating an initial random surface using fractal perturbation. The development patterns of such surfaces appear to provide realistic simulations of drainage initiation and development on relatively uniform surfaces. At first glance it appears a trifle ironic that Kirkby was forced to jump from a deterministic model to a probabilistic technique at this juncture. However, he had discussed the likelihood of this need and the reasons for it earlier in his work (Kirkby 1976a, pp. 3–4).

One important outcome of Kirkby's approach is that slope forms may be associated directly with dominant processes (Fig. 13.7). Such profiles develop independently of initial slope form; this implies that at some timescales slopes may be considered as time-independent. It also means that the model is suitable for forecasting slope evolution, but is much more restricted in its use for hindcasting (Kirkby 1976a, p. 5), and may be, in some circumstances, wholly unsuitable.

In conclusion, it is quite obvious that over the last 15 years Kirkby has developed incrementally a model that may now be described as comprehensive. It not only contains carefully developed submodels, but also appears to offer great promise because of its similarity to modeling in other disciplines, an attribute that makes interdisciplinary linkage attractive and feasible. Comparisons of analytical and numerical solutions of his equations also provide opportunities to assess the significance of initial assumptions. Finally, his more recent use of numerical techniques (e.g., Kirkby 1985) provides suitable methods for investigation of many different aspects of slope development.

The two models: a brief comparison

In comparing the modeling approaches of Ahnert and Kirkby it is appropriate to remember that Kirkby's most recent developments post-date the later work of Ahnert by nearly a decade. Therefore, their greater degree of sophistication is to be expected. It is also worth noting the manner in which both models are limited by their data requirements. This is not a flaw in these particular models, but merely a limitation of all mathematical models that are intended to function beyond a purely analytical level.

Both models are deterministic, but the mathematics of Ahnert's model are discrete while those of Kirkby's are continuous. The latter is more elegant mathematically, but the former is probably more accessible. Regardless of these niceties, either model is capable of producing quite satisfactory results, provided that the spatial and temporal increments are kept small. Initially, Ahnert's discrete approach made expansion to two dimensions much easier. Kirkby's initial expansion to two dimensions required imposition of some constraints on his model in order to permit the mathematics to function.

The two-dimensional models differ in at least one important manner. Ahnert's model derives the spatial variability that promotes a drainage network from structural variations, and does not address drainage development under relatively uniform conditions. On the other hand Kirkby's choice of a fractal perturbation, while marking a stark shift from deterministic to probabilistic techniques, is specifically aimed at developing a drainage network from a relatively uniform initial surface. Clearly, this is not a question of right or wrong, merely of difference, as both types of situation occur in the real world.

Kirkby's model appears to be stronger in its capacity to absorb detailed submodels as they are developed. Such refinements represent upgraded inputs to the main model. Kirkby's model also contains a stronger analytical framework than does Ahnert's, because of the early emphasis on analytical solutions. It is true that this probably slowed model development, as well as restricting its domain, but it has resulted in a more firmly based present product.

Even now neither model has really sound ways in which to handle either landsliding or lithologic variability. Of the two, landsliding probably represents less of a modeling problem than does lithologic variability. This is certainly a relative statement, because landsliding is extremely complex, but the enormous complexity of lithologic variability would seem likely to defy analytical solution in the foreseeable future.

In short, these two models should be seen in a largely sequential fashion. Ahnert's research represents a substantial and sustained effort to initiate and extend mathematical slope modeling. It has a heavy numerical emphasis because many facets of the topics involved were only available as a result of empirical work (again emphasizing the limiting nature of data requirements upon mathematical approaches to geomorphology). By adopting this approach Ahnert was able to provide early models of slope evolution that permitted efficient and useful ways in which to gain an understanding of the interactions between lithology, base level change, transport modes, and resulting hillslope forms. Kirkby's work represents a later generation of modeling and it is presently the state of the art. While it has benefited from its predecessor, Kirkby's model reaches further back than Ahnert's in the sense that it has a stronger analytical foundation. Another strength of Kirkby's model is the coupling of the model components and the ability of the overall model to absorb submodels as they are developed and/or refined. One of the most interesting facets of Kirkby's work at present is the emerging dilemma of where to make the transition from deterministic to probabilistic mathematical modeling. On purely *a priori* grounds analytical arguments should be taken as far as possible, and similarly deterministic techniques should be pursued until insurmountable difficulties dictate the need to pursue probabilistic (or stochastic) techniques. While this is a mathematical issue in the immediate context, it is also one more illustration of the complexities in geomorphology stemming from limited theoretical development and scale.

14 Diagnosis and prognosis

Collectively, geomorphologists appear to be of the opinion that landforms and landscapes change over time. This consensus is a distinctly limited one because it does not extend to the sequence(s) and rate(s) of landform and landscape change. In a similar vein the vast majority of geomorphologists would agree that present-day landforms and landscapes differ from each other – in at least some parts of the world. Again this represents a superficial consensus because there is no universally accepted explanation of why this is so and, if anything other than extreme cases were to be contrasted, agreement is unlikely and dispute probable.

All scientists are confronted by the need to choose an appropriate research objective at the beginning of a project. This step is not the focus of attention in this text and, accordingly, will not be treated at length here. However, the availability of diverse and often conflicting theoretical frameworks in geomorphology requires that the second step for a research geomorphologist commencing a project is to select a preferred theoretical standpoint. This entire text is founded on the notion that this second step is unavoidable and is therefore best made consciously rather than unwittingly. Consequently, this concluding chapter is devoted to highlighting some of the fundamental choices that must be made and to suggesting *a* (not *the*) sequence of steps that will provide relatively inexperienced researchers with an acceptable format with which to approach projects until they develop a better one for themselves.

Geomorphology: history or science?

There is a quite widely held opinion in academia that there is a fundamental methodological difference between history on the one hand and science on the other. The crux of this distinction is that history consists of a stream of unique events that as a result of their uniqueness are not amenable to generalization and, consequently, are unapproachable by the scientific method. The contrast is completed by viewing generalization and the construction of covering laws as the touchstone of the scientific method. This supposed dichotomy has also arisen with specific reference to geology and geomorphology: indeed, Brown's (1974, p. 456) remark, cited earlier, that "there is no science of singularities", while referring to geology specifically, represents an excellent generic summary of the situation.

Simpson (1963), Reynaud (1971), Twidale (1977), Kennedy (1980), and Thornes (1983) are among those who have discussed the supposed conflict between geomorphology as history and geomorphology as science. The pervasiveness of the historical component in geomorphology is apparent in Twidale's (1977, p. 88) words:

> The purpose of geomorphological research is to derive an explanatory account of the earth's surface. Geomorphology is above all historical in nature, in outlook and in approach. All landforms have evolved in the past – either yesterday, last week, last month, last year, a few thousand years ago, or, in the tropical lands in particular (though not exclusively), several scores or hundreds of millions of years ago – so that recourse has perforce to be made to past events and conditions in order to explain and understand the landscape.

The difficulty is not that Twidale's words misrepresent the objectives of geomorphologists, but that objectives themselves never prescribe theory and/or methodology. Therefore, the issue remains whether or not there is an inherent methodological distinction between history and science.

Simpson (1963) tried to resolve the conflict of geology (and implicitly geomorphology) as history or science by suggesting that there are really two types of science and that both are seen in geology. He (Simpson 1963, pp. 24–5) labeled the two characteristics of geology "configurational" and "immanent". He defined history as "configurational change through time, i.e., a sequence of real, individual but unrelated events". While he characterized "immanent" as: "The unchanging properties of matter and energy and the likewise unchanging processes and principles arising there-from are *immanent* in the material universe. They are nonhistorical, even though they occur and act in the course of history" [original emphasis]. While conceding that all sciences contain both configurational and imma-nent components, Simpson suggested that it is feasible to distinguish between nonhistorical sciences (e.g., chemistry) preoccupied with imma-nent issues and historical sciences (e.g., geology) dominated by historical themes. This kind of thinking may be rephrased (e.g., Gerrard 1984) as the difference between physical sciences that are focused upon "How" ques-tions that may be answered purely in terms of physical and chemical processes, and history that is focused upon "How come" questions that implicitly embrace a temporal axis.

Watson (1966, 1969) presented a strong rebuttal of Simpson's distinction between historical and nonhistorical sciences. The quintessential point in Watson's (1966) presentation was that all sciences contain both historical and nonhistorical components. He based this claim on the distinction between "*types* of events" (noting that these are abstractions) and "*particular* events which have specific coordinates in time and space" (Watson 1966, p. 175). As Watson pointed out, all sciences have types of events, which because they are nonspecific are subject to generalization, but all sciences also have particular events, which form the basis of the historical approach. Furthermore, he noted that ideas about the importance of types of events cannot be developed without reference to particular events and, conversely, evaluation of the potential importance of a particular event is actually dependent upon assessment of the type of event it represents. This led him to conclude that, while in some sciences there is a tendency to emphasize either types of events or particular events, the interdependence of the two categories actually precludes subdivision of sciences into historical and nonhistorical.

While Watson's (1966) argument is persuasive, it rather obviously destroys one possible line of reasoning by which geology (and again geomorphology by inference) may be justified as an independent science. The question may be posed: "Why not view geomorphology as the behavior of atoms and treat the entire situation as part of chemistry and/or physics?" This particular issue was addressed in a later paper by Watson.

Watson (1969, p. 489) pointed out that there are irreducible geological facts; for example, it is not feasible to discuss the development of a mountain range simply by describing the behavior of the constituent atoms because, while the latter may be technically correct with respect to chemistry and/or physics, something that is inherently "geology" (i.e., the concepts "mountains" and "mountain evolution") is lost in such a reduction. Obviously, a similar argument may be made with respect to geomorphology; in fact, such a case was made by Chorley (1978, p. 10), who presented it as a need for geomorphologists to coarsen their scale of interest.

If Watson's analysis is accepted, geology or geomorphology is either history or science not by virtue of changing the "facts" examined but by virtue of the theoretical standpoint from which the facts are examined. To view facts as a stream of unique events is to treat them historically, to generalize them is to treat them scientifically. Once again the significance, power, and limitations of theory emerge pre-eminent.

In the final analysis, the distinction between history and science really depends on whether an individual views the world from the perspective of a "lumper" (scientist) or "splitter" (historian). Indeed, even this is misleading because it labels the vast majority of historians in an entirely false manner. A better distinction is between a scientist (including many historians) and an undiscerning collector. However, once cast in the mold of a scientist, the geomorphologist must become preoccupied with theory as the only pathway leading to meaningful generalization.

What kinds of theory?

Thornes (1983) shed light on the sorts of theory that geomorphologists need and should be developing. In his paper Thornes distinguished between "theoretical" and "historical–inferential" approaches to geomorphology. "Theoretical" implicitly embraces notions of a falsifiable statement being made and subsequently tested, while "historical–inferential" implies an interest in the sequence of events, and perhaps the belief that history tends to repeat itself, but lacks any formally defined rules.

He (Thornes 1983) pointed out that geomorphologists may map, or display, landscapes as regions (domains) where particular processes are dominant. The core of process geomorphology has been recognition of such regions and, in turn, is founded on equilibrium concepts (i.e., between the dominant process(es) and the landform(s)). If landscapes are displayed in this fashion, there are obviously boundaries between the regions: furthermore, if there is a shift in dominant process(es), a landform or landscape will be driven from one region across a boundary into another region. Such a transition would be associated with important morphological changes.

Completion of the above set of ideas may be made by considering what circumstances initiate and develop the structure or distribution of all the regions or domains, a task dubbed "evolutionary geomorphology" by Thornes (1983, p. 227), and also by studying the pathway of a particular landscape as it moves through various process regions or domains over time. The latter is the traditional subject matter of denudation chronology (Ch. 3).

Consequently, the theoretical needs in geomorphology that are implied by Thornes's analysis are: (1) formal statements addressing landscape behavior while located within a domain; and (2) formal statements addressing landscape behavior while crossing a boundary between domains. The former has often been addressed in the recent past using the concepts of steady state or dynamic equilibrium. The second issue, that of landscape behavior while crossing domain boundaries, has not been treated theoretically, as opposed to historical–inferentially, since the demise of the Davisian model.

Schumm's (e.g., Schumm 1979) introduction of dynamic metastable equilibrium, and particularly the concepts of intrinsic thresholds and complex responses (Ch. 6), represented a significant step forward in geomorphic theory. Aside from the specifics of the concepts. Schumm's ideas paved the way in geomorphology for recognition and theoretical consideration of much more complicated behavioral patterns than those associated with steady state and dynamic equilibrium (Schumm 1985). Fortunately, interest in the behavior of systems that are not in equilibrium had moved to the forefront in some other disciplines and work by individuals such as Prigogine (1980) and Thom (1975) introduced theoretical frameworks that could be applied to unstable situations.

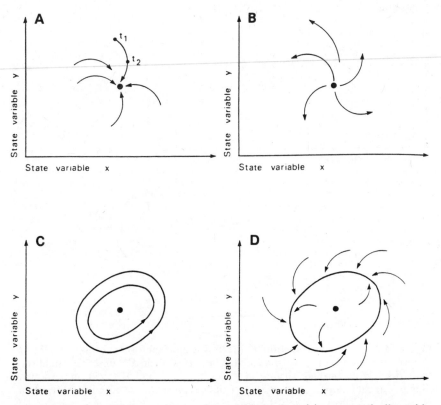

Figure 14.1 Some different forms of system behavior: (a) asymptotically stable (attractor); (b) unstable; (c) neutrally stable; (d) stable limit cycle. (From Thornes (1983).)

Figure 14.1a shows a situation in which a system always tends toward a characteristic relationship, while Figure 14.1b shows one that is inherently unstable and consequently self-destructive (i.e., dominated by positive feedback). Figures 14.1c and 14.1d show two different types of oscillation; in either case the system circles about some mean value without ever attaining it. Thornes (1983) discussed several other kinds of developmental behavior, among the more well-known of which is that associated with Thom's (1975) catastrophe theory (Fig. 14.2). These ideas have only been pursued occasionally in geomorphology (e.g., Graf 1979). The gist of this concept is that there is a three-dimensional surface upon which the system is stable. However, should change occur, there may be very rapid change (called a catastrophe) as the system jumps from one position to another on the equilibrium surface. It is worth emphasizing that within geomorphology catastrophe theory implies that more than one landform might be stable within any specific context.

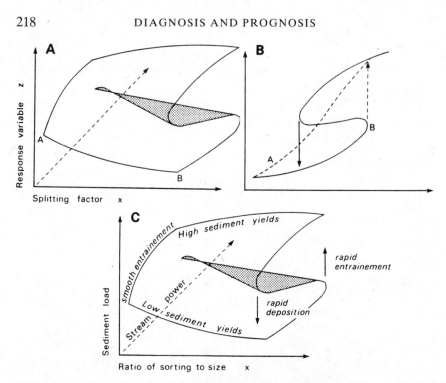

Figure 14.2   The cusp catastrophe; (a) basic canonical form showing equilibrium surface for response variable (shaded area on surface in area of unstable equilibria); (b) cross section through surface at A (dotted line) and B (solid line) showing "jumps" or catastrophes; (c) cusp catastrophe representation of sediment transport process. (from Thornes (1983).)

Clearly, the concepts shown schematically in Figures 14.1 and 14.2 represent theoretical evolutionary geomorphology. In other words, they offer formal (falsifiable) statements about the behavior of landforms or landscapes through time, including explanations of why a landform or landscape may fall within a particular domain and/or move from one domain to another. They afford a similar theoretical foundation for denudation chronology.

In summary, geomorphologists need to put the development of a body of theory addressing landform and landscape development over all timespans at the top of their collective agenda. Such a step implies that they must leave behind the equilibrium concepts that dominated process geomorphology during the 1960s and 1970s, and stemmed in large part from renewed interest in the much earlier work of G. K. Gilbert (Ch. 11). If such a task becomes paramount, and it should, there will also have to be a substantive effort to create formal theory that addresses spatial issues. In this latter task geomorphologists may still draw heavily upon equilibrium concepts, which implicitly contain the notion that spatial, rather than temporal, factors are dominant. As spatial factors pollute evolutionary geomorphology, so temporal factors pollute equilibrium geomorphology. Consequently, theoretical development in either sphere is ultimately limited by theoretical development in the other one: if theoretical development in the two spheres is badly mismatched, there is no adequate basis for discerning between noise and signal in either sphere.

Concentration upon these two themes is not a second-rate, or poor-man's, intellectual endeavor; nor will it lead to creation of a second-rate discipline. While it is true that these are issues that have been investigated in other sciences, they have not generally been focal. Such a theoretical corpus would inevitably have to address scale linkage (both temporal and spatial), which still looms as the great specter over geomorphology. In short, there is a significant and worthwhile theoretical niche that geo-morphologists could occupy and develop within the spectrum of science. Furthermore, geomorphologists are not bereft of suitable starting points: there is a wealth of unexploited theoretical, or potentially theoretical (i.e., they could be restated formally), suggestions in the literature (e.g., Crick-may's (1959, 1975) hypothesis of unequal activity and Garner's (1959) ideas concerning climatic shifts and Andean development – to cite just two examples that come to mind. Interestingly enough, a few eccentric indi-viduals in the discipline have been preoccupied with theoretical develop-ment and pursued it consistently and outspokenly: this exotic behavior has earnt them great eminence! There ought to be a lesson in this seeming paradox.

The yellow brick road

The avowed purpose of this text was to introduce students to the theoretical content of contemporary geomorphology. To this point the task has been attempted by reviewing elemental components of theoretical geo-morphology, as well as some of the better-known derivative composites or structures. Reviews are useful undertakings; they invariably teach their creator a great deal and offer their readers a little enlightenment. However, reviews are also timid creations, bereft of the zest that judgment offers. Accordingly, this last section is a profession of geomorphology, written in living proof of the colloquial American aphorism – no pain, no gain!

Objective(s)

Theory is inevitably conditioned by the purpose(s) or objective(s) for which it was created. This reality may be interpreted broadly or narrowly and, clearly, this text has been based on a broad interpretation. In addressing objectives in geomorphology, issues that have been purposely pushed into the background are being brought into the limelight. Even here they will be addressed only in the most generic terms.

Geomorphologists must have development of covering laws as their primary goal. Using Watson's (1966) terminology, this means that geomorphologists must view all events primarily in the context of "types of events". As Watson pointed out, this may be extremely difficult, as geologists and geomorphologists are frequently confronted by events that are infrequent. Consequently, it is often difficult to test geomorphic laws. Whatever the magnitude of this difficulty, it is important to avoid the (often appealing) trap of merely recounting events as a sequence or series of unique occurrences. If geomorphology is threatened by the enormous growth of Quaternary science, and to some degree this does appear to be a danger, it is primarily because Quaternary science tends to emphasize the role of the "historical–inferential" approach over that of the theoretical.

The second objective that geomorphologists must retain is a preoccupation with geomorphology. As Watson (1969) and Chorley (1978) have suggested, there is subject matter that is irreducibly geomorphology. This is not to suggest that all microscale work is to be proscribed or taboo. However, when a researcher conducts work at a scale at which there are no irreducible geomorphic components, it behooves him or her to demonstrate the linkage that ties such work to irreducible geomorphic components and the relevance of his or her work to their behavior. It is not sufficient merely to note that landforms and landscapes are composed of atoms.

Methodology

Methodology is the product of theory, and consequently, a discussion of the appropriate manner in which to pursue geomorphic research (i.e., methodology) is implicitly a discussion of geomorphic theory. Here it will be presented as a series of steps.

Step one: Develop, maintain, and sustain a strong personal knowledge of geomorphic theory.

Virtually every student of geomorphology receives constant encouragement to expand his or her field experience. This must be complemented by an equally enthusiastic pursuit of theoretical knowledge. Fieldwork and theory are more than complementary, they are symbiotic; one without the other is a nonentity.

Step two: Formulate research questions that are firmly imbedded in geomorphic theory and pose them in a falsifiable fashion.

The objective is technically a matter of personal choice. However, it is likely to be at least guided (if not dictated) by peer pressure, existing literature, etc. The three fundamental factors in geomorphology are space, time, and their interaction. If it is presumed that a question can only be answered in terms of one of them, the issue has been prejudged (remember Davis's answer to graded rivers, versus older and newer explanations).

Step three: Build and attempt to falsify a characteristic-form hypothesis.

All geomorphic research is ultimately based on the decision to view landforms or landscapes as equilibrium or relaxation forms (a decision that is nearly always a scale-dependent one). However, as it is scientifically appropriate to pursue the simplest approach that is theoretically acceptable, a characteristic-form approach should be attempted before a relaxation-form one. Characteristic-form concepts are inherently simpler than relaxation-form ones because they lack the temporal component and associated difficulties.

Characteristic-form concepts are invariably associated with equilibrium (see Davidson (1978), Chorley and Kennedy (1971), Graf (1979), Thornes (1983), and Huggett (1985) for appropriate starting points). In geomorphology, equilibrium is applied to form and, because of the difficulty of measuring energy inputs and outputs (plus the inefficiency of most natural systems), it seems that geomorphologists will continue to pursue equilibrium concepts in the near future only to the extent that they can be applied to the behavior of form over time.

The variety of equilibrium concepts is large and growing, as has been shown. Fundamentally, the newer ideas are underpinned by the notion that landforms may oscillate about some preferred form, without ever actually achieving it. This complicates interpretation because short-term form change may be characteristic of both stability and instability, and the two may be difficult to distinguish. More recent versions of equilibrium also incorporate multiple equilibrium positions, implying that equilibrium may be represented by different landforms.

Present evidence suggests that equilibrium or time-independent models will be unsatisfactory in many instances. Consequently, if such a model produces unsatisfactory results, it is necessary to move onto the next step.

Step four: Build and attempt to falsify a relaxation-form hypothesis.

Falsification of a characteristic-form hypothesis implies that it is appropriate to develop an evolutionary hypothesis of the landform's or landscape's behavior. As Thornes (1983) pointed out, such theory has not been prevalent in geomorphology since the downfall of the Davisian model. However, Thornes's (1983) own paper, as well as papers such as those of Knox (1972), Graf (1977, 1982), and Schumm (1979), all offer potentially quantifiable and, therefore, falsifiable concepts of landform behavior over time.

It is possible that development of better evolutionary theory will come from more complex models. However, it seems much more appropriate to approach the relevant issues by testing simple concepts that are carefully demarcated and rich in components with sound operational definitions. To return to two earlier themes, geomorphologists need to pursue Nystuen's

(1963) concept of abstraction (or at least simplification) (Ch. 7); this should be complemented by extensive development of experimental fieldwork as suggested by Church (1984) (Ch. 9) because this affords the necessary means of testing. More complex models may then be developed from well-established components, rather than from a myriad of uncertainties.

Some footnotes

Multiple working hypothesis

The four steps suggested above are not intended to represent an alternative to the multiple-working-hypothesis approach to research, rather they are a complement. At each of the suggested steps the multiple-working-hypothesis approach is perfectly applicable.

On linguistic preference

The emphasis placed herein upon theory building implicitly favors the precision inherent to mathematics. Within mathematics itself, theory building similarly favors deterministic over probabilistic approaches, as well as analytical ones over numerical ones. To favor is not the same as to mandate, and that which is not favored is not proscribed.

The real test when choosing an appropriate language is the objective not the language. Given the objective, the language selected should be that which permits it to be expressed and tested most precisely. There are many instances where it is not feasible to articulate mathematically an objective, but this does not make the objective worthless – it makes mathematics inappropriate.

Conversely, geomorphologists could advance their discipline significantly if they would, as a group, make the extra effort to express their concepts more precisely. Certainly, the easy way to do this is to become more mathematical. Changing languages will not be a universal panacea, and does not even insure progress, but it will have the tendency to force greater rigor upon the research community.

Even simple abstraction and conceptualization such as Jenny's (1941) soil state factor equation and Peltier's (1975) general landform equation represent positive steps. As even some simple nonlinear equations can produce complex results (Thornes 1983), there is no reason to believe that a geomorphologist must be an outstanding mathematician in order to contribute. Confirmation of this suggestion is found in the many theoretical contributions that are currently held in high esteem, but are simple mathematically. A massive step forward could be made by universal use of simple analytical mathematical expressions.

There is another reason to favor mathematics. All geomorphology is modeling, and all geomorphology spans some sort of time period. If geomorphologists restrict themselves to timespans over which they can make direct observations, geomorphology will be enormously curtailed. Conversely, if geomorphologists expand their theories to embrace the distant past and/or the future, they have few options but to build mathematical models and test them mathematically. The options are, in fact, very unappealing, the most obvious being untested theory (very dangerous stuff!), untestable theory (equally dangerous), and no theory (intellectual suicide).

Conclusion

Geomorphology, like all human endeavor, is not only pervaded by theory, it is the creation of theory. While most students are exhilarated by strenuous field trips, it has been the purpose of this text to suggest that geomorphology offers even more challenges, and therefore even greater exhilaration, intellectually. Furthermore, a balanced combination of theory and fieldwork yields a veritable Buddhist's nirvana!

Ultimately there are really only two points to be made in summarizing what has been attempted here. First, theory is inescapable; therefore it is better to master it than to have it baffle, befuddle, and bludgeon one. Second, theory is both a lodestar (a guiding ideal) and a touchstone (a test for determining genuineness), whenever two such valuable possessions may be had for the same price – seize the opportunity!

References

Abler, R., J. S. Adams & P. Gould 1971. *Spatial organization: The geographer's view of the world*. Englewood Cliffs, NJ: Prentice-Hall.

Abrahams, A. D. (ed.) 1986. *Hillslope processes*. Binghamton Symp. Geomorph., Int. Ser., no. 16. London: Allen & Unwin.

Ackoff, R. L. (with S. K. Gupta & J. S. Minas) 1962. *Scientific method: optimizing applied research decisions*. New York: Wiley.

Ager, D. V. 1981. *The nature of the stratigraphical record*, 2nd edn. London: Macmillan.

Ager, D. V. 1984. The proper study of geology is rocks. *New scientist* **1414**, 42.

Agnew, C. T. 1984. Checkland's soft systems approach – a methodology for geographers? *Area* **16**, 167–74.

Ahnert, F. 1967. The role of the equilibrium concept in the interpretation of landforms of fluvial erosion and deposition. In *L'evolution des versants*, P. Macar (ed.), 23–41. Liège: University of Liège.

Ahnert, F. 1970. Functional relationships between denudation, relief, and uplift in large mid-latitude drainage basins. *Am. J. Sci.* **268**, 243–63.

Ahnert, F. 1971. *A general and comprehensive theoretical model of slope profile development*. Univ. Maryland, Occas. Pap. Geog., no. 1.

Ahnert, F. 1973. *COSLOP 2 – a comprehensive model program for simulating slope profile development*. Geocom Bull. **6**, 99–122.

Ahnert, F. 1976. Brief description of a comprehensive three-dimensional process–response model of landform development. *Z. Geomorph, N.F. Suppl.* **25**, 29–49.

Ahnert, F. 1977. Some comments on the quantitative formulation of geomorphological processes in a theoretical model. *Earth Surf. Processes* **2**, 191–201.

Ahnert, F. 1980. A note on measurements and experiments in geomorphology. *Z. Geomorph. N.F. Suppl.* **35**, 1–10.

Allen, J. R. L. 1974. Reaction, relaxation and lag in natural sedimentary systems: general principles, examples and lessons. *Earth Sci. Rev.* **10**, 263–342.

Amerman, C. R. 1965. The use of unit-source watershed data for runoff prediction. *Water Resources Res.* **1**, 499–507.

Andersson, J. G. 1906. Solifluction, a component of sub-aerial denudation, *J. Geol.* **14**, 91–112.

Andrews, J. T. 1971. *Techniques of till fabric analysis*. Br. Geomorph. Res. Group Tech. Bull., no. 6.

Baker, V. R. & S. Pyne 1978. G. K. Gilbert and modern geomorphology. *Am. J. Sci.* **278**, 97–123.

Bakker, J. P. & J. W. N. LeHeux 1946. Projective-geometric treatment of O. Lehmann's theory of the transformation of steep mountain slopes. *K. Ned. Akad. Wet. B* **49**, 533–47.

Bakker, J. P. & J. W. N. LeHeux 1947. Theory of central rectilinear recession of slopes I and II. *K. Ned. Akad. Wet. B* **50**, 959–66 and 1154–62.

Bakker, J. P. & J. W. N. LeHeux 1950. Theory of central rectilinear recession of slopes III and IV. *K. Ned. Akad. Wet. B* **53**, 1073–84 and 1364–74.

Bakker, J. P. & J. W. N. LeHeux 1952. A remarkable new geomorphological law. *K. Ned. Akad. Wet. B* **55**, 399–410 and 554–71.

Barrett, P. J. 1980. The shape of rock particles, a critical review. *Sedimentology* **27**, 291–303.

Bates, R. L. & J. A. Jackson 1980. *Glossary of geology*, 2nd edn. Falls Church, VA: American Geological Institute.

Beckinsale, R. P. 1976. The international influence of William Morris Davis. *Geog. Rev.* **66**, 448–66.

Benedict, J. B. 1970. Downslope soil movement in a Colorado alpine region: rates, processes and climatic significance. *Arctic Alpine Res.* **2**, 165–226.

Bennett, R. J. & R. J. Chorley 1978. Environmental systems. London: Methuen.

Bird, J. H. 1975. Methodological implications for geography from the philosophy of K. R. Popper. *Scott. Geog. Mag.* **91**, 153–63.

Birkeland, P. W. 1984. *Soils and geomorphology*. New York: Oxford University Press.

Blair, D. J. & T. H. Bliss 1967. *The measurement of shape in geography*. Quant. Bull. Dept Geog., Univ. Nottingham, no. 11.

Boch, S. G. 1946. Snow patches and snow erosion in the northern part of the Urals. *Vses. Geog. Obshch. Izv.* **78**, 207–22.

Boch, S. G. 1948. Some remarks on the nature of snow erosion. *Vses. Geog. Obshch. Izv.* **80**, 609–11.

Bovis, M. J. & C. E. Thorn 1981. Soil loss variation within a Colorado alpine area. *Earth Surf. Processes* **6**, 151–63.

Box, G. E. P. & G. M. Jenkins 1976. *Time series analysis: forecasting and control*, 2nd edn. New York: Holden-Day.

Boyce, R. B. & W. A. V. Clark 1964. The concept of shape in geography. *Geog. Rev.* **54**, 561–72.

Boyer, P. S. 1979. Peneplain vs peneplane: a correspondence. *J. Geol. Educ.* **27**, 59–63.

Bradley, R. S. 1985. *Quaternary paleoclimatology*. Boston: Allen & Unwin.

Bremer, H. 1983. Albrecht Penck (1858–1945) and Walter Penck (1888–1923), two German geomorphologists. *Z. Geomorph. N.F.* **27**, 129–38.

Bretz, J. H. 1923. Channeled scablands of the Columbia Plateau. *J. Geol.* **31**, 617–49.

Brice, J. C. 1964. *Channel patterns and terraces of the Loup River in Nebraska*. U.S. Geol. Surv. Prof. Pap., no. 422–D.

Brown, B. W. 1974. Induction, deduction, and irrationality in geologic reasoning. *Geology* **2**, 456.

Brown, J. R. 1976. *Ergodic theory and topological dynamics*. New York: Academic Press.

Brunsden, D. & R. H. Kesel 1973. Slope development on a Mississippi River bluff in historic time. *J. Geol.* **81**, 576–98.

Brunsden, D. & J. B. Thornes 1979. Landscape sensitivity and change. *Trans. Inst. Br. Geog. N.S.* **4**, 463–84.

Bryan, R. & A. Yair (eds) 1982. *Badland geomorphology and piping*. Norwich: Geo Books.

Büdel, J. 1982. *Climatic geomorphology*. Princeton, NJ: Princeton University Press.

Bull, W. B. 1975a. Allometric change of landforms. *Bull. Geol. Soc. Am.* **86**, 1489–98.

Bull, W. B. 1975b. Landforms that do not tend toward a steady state. In *Theories of landform development*, W. N. Melhorn & R. C. Flemal (eds), 111–28. Boston: Allen & Unwin.

Bull, W. B. 1977. Allometric change of landforms: reply. *Bull. Geol. Soc. Am.* **88**, 1200–2.

Burns, S. F. & P. J. Tonkin 1982. Soil-geomorphic models and the spatial distribution and development of alpine soils. In *Space and time in geomorphology*, C. E. Thorn (ed.) 25–43. Binghamton Symp. Geomorph., Int. Ser. no. 12. London: Allen & Unwin.

Caine, N. 1968. The log-normal distribution and rates of soil movement: an example. *Rev. Géomorph. Dyn.* **43**, 1–7.

Caine, N. 1974. The geomorphic processes of the alpine environment. In *Arctic and alpine environments*, J. D. Ives & R. G. Barry (eds). London: Methuen.

Caine, N. 1976. A uniform measure of subaerial erosion. *Bull. Geol. Soc. Am.* **87**, 137–40.

Caine, N. 1979. The problem of spatial scale in the study of contemporary geomorphic activity on mountain slopes (with special reference to the San Juan Mountains). *Stud. Geomorphol. Carpatho-Balcanica* **13**, 5–22.

Caine, N. 1982. The spatial variability of surficial soil movement rates in alpine environments. In *Space and time in geomorphology*, C. E. Thorn (ed.), 45–57. Binghamton Symp. Geomorph. Int. Ser., no. 12. London: Allen & Unwin.

Carroll, J. B. 1966. Introduction. In *Language, thought and reality: selected writings of Benjamin Lee Whorf*, J. B. Carroll (ed.), 1–34. Cambridge, MA: MIT Press.

Carson, M. A. 1967. *The evolution of straight debris-mantled hillslopes*. Ph.D. Dissertation. Cambridge University.

Carson, M. A. 1971. *The mechanics of erosion*. London: Pion.

Carson, M. A. & M. J. Kirkby 1972. *Hillslope form and process*. Cambridge: Cambridge University Press.

Carter, C. S. & R. J. Chorley 1961. Early slope development in an expanding stream system. *Geol. Mag.* **98**, 117–30.

Caws, P. 1965. *The philosophy of science*. Princeton, NJ: Van Nostrand.

Chamberlin, T. C. 1897. The method of multiple working hypotheses. *J. Geol.* **5**, 837–48.

Checkland, P. B. 1972. Towards a systems-based methodology for real-world problem solving. *J. Syst. Eng.* **3**, 87–116.

Chorley, R. J. 1962. *Geomorphology and general systems theory*. U.S. Geol. Surv. Prof. Pap., no. 500–B.

Chorley, R. J. 1963. Diastrophic background to twentieth century geomorphological thought. *Bull Geol. Soc. Am.* **74**, 953–70.

Chorley, R. J. 1965. A re-evaluation of the geomorphic system of W. M. Davis. In *Frontiers in geographical teaching*, R. J. Chorley & P. Haggett (eds), 21–38. London: Methuen.

Chorley, R. J. 1967. Models in geomorphology. In *Models in geography*, R. J. Chorley & P. Haggett (eds), 59–96. London: Methuen.

Chorley, R. J. 1972. Spatial analysis in geomorphology. In *Spatial analysis in geomorphology*, R. J. Chorley (ed.), 3–16. London: Harper & Row.

Chorley, R. J. 1978. Bases for theory in geomorphology. In *Geomorphology: present problems and future prospects*, C. Embleton, D. Brunsden, & D. K. C. Jones (eds), 1–13. Oxford: Oxford University Press.

Chorley, R. J. & R. P. Beckinsale 1980. G. K. Gilbert's geomorphology. In *The scientific ideas of G. K. Gilbert*, E. L. Yochelson (ed.). Geol. Soc. Am. Spec. Pap., no. 183.

Chorley, R. J. & L. S. D. Morley 1959. A simplified approximation for the hypsometric integral. *J. Geol.* **67**, 566–71.

Chorley, R. J. & B. A. Kennedy 1971. *Physical geography: a systems approach*. London: Prentice-Hall.

Chorley, R. J. & R. P. Beckinsale & A. J. Dunn 1973. *The history of the study of landforms*. Vol. 2: *The Life and Work of William Morris Davis*. London: Methuen.

Chorley, R. J., A. J. Dunn & R. P. Beckinsale (eds) 1964. *The history of the study of landforms or the development of geomorphology*. Vol. 1. London: Methuen.

Chorley, R. J., D. E. G. Malm & H. A. Pogorzelski 1957. A new standard for estimating drainage basin shape. *Am. J. Sci.* **255**, 138–41.

Chorley, R. J., S. A. Schumm & D. E. Sugden 1984. *Geomorphology*. London: Methuen.

Christian, C. S. & G. A. Stewart 1953. *General report on survey of Katherine–Darwin region, 1946*. C.S.I.R.O. Aust. Land Res. Ser., no. 1.

Church, M. 1980. Records of recent geomorphological events. In *Timescales in geomorphology*, R. A. Cullingford, D. A. Davidson & J. Lewin (eds), 13–29. Chichester: Wiley.

Church, M. 1983. "Space and time in geomorphology", C. E. Thorn (ed.), Book review. *Earth Surf. Processes Landforms* **8**, 514.

Church, M. 1984. On experimental method in geomorphology. In *Catchment experiments in fluvial geomorphology*, T. P. Burt & D. E. Walling (eds), 563–80. Norwich: Geo Books.

Church, M. & D. M. Mark 1980. On size and scale in geomorphology. Prog. Phys. Geog. **4**, 342–90.

Church, M. & J. M. Ryder 1972. Paraglacial sedimentation: consideration of fluvial processes conditioned by glaciation. *Bull. Geol. Soc. Am.* **83**, 3059–72.

Ciolkosz, E. J. 1978. Periglacial features of Pennsylvania. In *Quaternary deposits and soils of the central Susquehanna Valley of Pennsylvania*, 35–8. Guidebook 41st Annu. Reunion Northeast Friends of the Pleistocene.

Cole, J. P. & C. A. M. King 1968. *Quantitative geography*. London: Wiley.

Collins, L. 1975. *An Introduction to Markov chain analysis*. Concepts Tech. Mod. Geog., no. 1. Norwich: Geo Abstracts.

Costa, J. E. 1975. Effects of agriculture on erosion and sedimentation in the Piedmont province, Maryland. *Bull. Geol. Soc. Am.* **86**, 1281–6.

Cox, N. J. 1977. Allometric change of landforms: discussion. *Bull. Geol. Soc. Am.* **88**, 1199–200.

Craig, R. G. 1982a. Criteria for constructing optimal digital terrain models. In *Applied geomorphology*, R. G. Craig & J. L. Craft (eds), 108–30. Binghamton Symp. Geomorph., Int. Ser., no. 11. London: Allen & Unwin.

Craig, R. G. 1982b. The ergodic principle in erosion models. In *Space and time in geomorphology*, C. E. Thorn (ed.), 81–115. Binghamton Symp. Geomorph. Int. Ser., no. 12. London: Allen & Unwin.

Craig, R. G. & M. L. Labovitz 1981. *Future trends in geomathematics*. London: Pion.

Crickmay, C. H. 1959. *A preliminary enquiry into the formulation and applicability of the geological principal of uniformity*, Calgary: Crickmay.

Crickmay, C. H. 1975. The hypothesis of unequal activity. In *Theories of landform development*, W. N. Melhorn & R. C. Flemal (eds), 103–9. Boston: Allen & Unwin.

Culling, W. E. H. 1960. Analytical theory of erosion. *J. Geol.* **68**, 336–44.

Culling, W. E. H. 1963. Soil creep and the development of hillside slopes. *J. Geol.* **71**, 127–61.

Culling, W. E. H. 1965. Theory of erosion on soil-covered slopes. *J. Geol.* **73**, 230–55.

Czech, H. & K. C. Boswell (transl.) 1972. *Morphological analysis of land forms*. New York: Hafner. (This is a translation of Walter Penck's original text. A citation for the original is given below. First edition of translation published in 1953.)

Dacey, M. F. 1973. Some questions about spatial distributions. In *Directions in geography*, R. J. Chorley (ed.), 127–51. London: Methuen.

Darwin, C. R. 1869. *On the origin of species by means of natural selection; or, the preservation of favoured races in the struggle for life*, 5th edn. London: John Murray. (First edition published in 1859.)

Davidson, D. A. 1978. *Science for physical geographers*. London: Edward Arnold.

Davis, J. C. 1973. *Statistics and data analysis in geology*. New York: Wiley.

Davis, W. M. 1885. Geographic classification, illustrated by a study of plains, plateaus, and their derivatives. *Proc. Am. Assoc. Adv. Sci.* **33**, 428–32.

Davis, W. M. 1892. The convex profile of bad-land divides. *Science* **20**, 245.

Davis, W. M. 1899. The geographical cycle. *Geog. J.* **14**, 481–504.

Davis, W. M. 1902. Base level, grade, and peneplain. *J. Geol.* **10**, 77–111.

Davis, W. M. 1905. The geographical cycle in an arid climate. *J. Geol.* **13**, 381–407.

Davis, W. M. 1909a. *Geographical essays*, D. W. Johnson (ed.). Boston: Ginn.

Davis, W. M. 1909b. The systematic description of land forms. *Geog. J.* **34**, 300–18, discussion 318–26.

Davis, W. M. 1912. *Die Erklärende Beschreibung der Landformen*, A. Ruhl (transl.). Leipzig: Teubner.

Davis, W. M. 1932. Piedmont benchlands and Primärrümpfe. *Bull. Geol. Soc. Am.* **43**, 399–440.

Demek, J. (ed.) 1972. *Manual of detailed geomorphological mapping*. Prague: Academia.

Denny, C. S. 1956. *Surficial geology and geomorphology of Potter County, Pennsylvania*. U.S. Geol. Surv. prof. Pap., no. 288.

DePloey, J. H. 1972. Enkele bevindingen befreffend erosieprocesses en hellingsevolutie op zandig substraat. *Tijdschr. Belg. Ver. Aardrijksk. Stud.* **41**, 43–67.

Derbyshire, E. (ed.) 1973. *Climatic geomorphology*. London: Macmillan.

Derbyshire, E. (ed.) 1976. *Geomorphology and climate*. London: Wiley.

Doornkamp, J. C. & C. A. M. King 1971. *Numerical analysis in geomorphology: an introduction*. London: Edward Arnold.

Dunne, T. & L. B. Leopold 1978. *Water in environmental planning*. San Francisco: W. H. Freeman.

Dury, G. H. 1964. *Principles of underfit streams*. U.S. Geol. Surv. Prof. Pap., no. 452–A.

Dury, G. H. 1965. *Theoretical implications of underfit streams*. U.S. Geol. Surv. Prof. Pap., no. 452–C.

Dury, G. H. 1966. The concept of grade. In *Essays in geomorphology*, G. H. Dury (ed.), 211–33. London: Heinemann.

Dury, G. H. 1969. *Perspectives on geomorphic processes*. Assoc. Am. Geog. Resource Pap., no. 3.

Dury, G. H. 1975. Neocatastrophism? *An. Acad. Cienc. Bras. (Supl.)* **47**, 135–51.

Dury, G. H. 1980. Neocatastrophism? A further look. *Prog. Phys. Geog.* **4**, 391–413.

Dury, G. H. 1981. *An introduction to environmental systems*. London: Heinemann.

Dutton, C. E. 1889. On some of the greater problems of physical geology. *Phil. Soc. Washington Bull.* **11**, 51–64.

Dylik, J. 1967. Solifluxion, congelifluxion and related slope processes. *Geog. Ann.* **49A**, 167–77.

Ebdon, D. 1978. *Statistics in geography: a practical approach*. Oxford: Blackwell.

Ehrlich, R. & B. Weinberg 1970. An exact method for characterization of grain shape. *J. Sed. Petrol* **40**, 205–12.

Embleton, C. & C. A. M. King 1975. *Periglacial geomorphology*, 2nd edn. New York: Wiley.

Evans, I. S. 1972. General geomorphometry, derivations of altitude and descriptive statistics. In *Spatial analysis in geomorphology*, R. J. Chorley (ed.), 17–90. London: Harper & Row.

Evans, I. S. & N. Cox 1974. Geomorphometry and the operational definition of cirques. *Area* **6**, 150–3.

Eyton, J. R. 1974. *An analytical surfacing methodology for application to the geometric measurement and analysis of landforms*. Unpublished Ph.D. Dissertation, University of Illinois at Urbana-Champaign.

Fahnestock, R. K. 1963. *Morphology and hydrology of a glacial stream – White River, Mount Rainier, Washington*. U.S. Geol. Surv. Prof. Pap., no. 422–A.

Fair, T. J. D. 1947. Slope form and development in the Interior of Natal, South Africa. *Geol. Soc. S. Afr. Trans.* **50**, 105–18.

Fair, T. J. D. 1948a. Hillslopes and pediments of the semi-arid Karroo. *S. Afr. Geog. J.* **30**, 71–9.

Fair, T. J. D. 1948b. Slope form and development in the coastal hinterland of Natal. *Geol. Soc. S. Afr. Trans.* **51**, 33–47.

Fairbridge, R. W. 1968. *The encyclopedia of geomorphology*. New York: Reinhold.

Fenneman, N. M. 1931. *Physiography of western United States*. New York: McGraw-Hill.

Fenneman, N. M. 1938. *Physiography of eastern United States*. New York: McGraw-Hill.

Ferguson, R. I. 1973. Regular meander path models. *Water Resources Res.* **9**, 1079–86.

Ferguson, R. I. 1976. Disturbed periodic model for river meanders. *Earth Surf. Processes* **1**, 337–47.

Flemal, R. C. 1971. The attack on the Davisian system of geomorphology: a synopsis. *J. Geol. Educ.* **19**, 3–13.

Flint, R. F., J. E. Sanders & J. Rodgers 1960. Diamictite, a substitute term for symmictite. *Bull. Geol. Soc. Am.* **71**, 1809.

Gardiner, V. 1974. *Drainage basin morphometry*. Br. Geomorph. Res. Group, Tech. Bull., no. 14.

Gardiner, V. & C. C. Park 1978. Drainage basin morphometry: review and assessment. *Prog. Phys. Geog.* **2**, 1–35.

Gardner, J. S. 1980. Frequency, magnitude, and spatial distribution of mountain rockfalls and rockslides in the Highwood Pass area, Alberta, Canada. In *Thresholds in geomorphology*, D. R. Coates & J. D. Vitek (eds), 267–95. Binghamton Symp. Geomorph., Int. Ser., no. 9. London: Allen & Unwin.

Gardner, R. & Scoging, H. (eds) 1983. *Mega-geomorphology*. Oxford: Oxford University Press.

Garner, H. F. 1959. Stratigraphic–sedimentary significance of contemporary climate and relief in four regions of the Andes mountains. *Bull. Geol. Soc. Am.* **70**, 1327–68.

Garner, H. F. 1974. *The origin of landscapes*. New York: Oxford University Press.

Gerrard, A. J. 1984. Multiple working hypotheses and equifinality in geomorphology: comments on the recent article by Haines-Young and Petch. *Trans. Inst. Br. Geog. N.S.* **9**, 364–6.

Getis, A. 1977. On the use of the term "random" in spatial analysis. *Prof. Geog.* **29**, 59–61.

Getis, A. & B. Boots 1978. *Models of spatial processes. An approach to the study of point, line and area patterns*. Cambridge: Cambridge University Press.

Gilbert, G. K. 1877. Report on the geology of the Henry Mountains. In *U.S. Geographical and Geological Survey of the Rocky Mountain Region*.

Gilbert, G. K. 1886. The inculcation of scientific method by example. *Am. J. Science* **31**, 284–99.

Gilbert, G. K. 1890. *Lake Bonneville*. U.S. Geol. Monogr., no. 1.

Gilbert, G. K. 1909. The convexity of hilltops, *J. Geol.* **17**, 344–51.

Gilbert, G. K. 1914. *The transportation of debris by running water*. U.S. Geol. Surv. Prof. Pap., no. 86.

Glen, J. W. 1952. Experiments on the deformation of ice. *J. Glaciol.* **2**, 111–14.

Goodman, N. 1967. Uniformity and simplicity. In *Uniformity and simplicity*, C. C. Albritton, Jr. (ed.), 93–9. Geol. Soc. Am. Spec. Pap., no. 89.

Gould, S. J. 1966. Allometry and size in ontogeny and phylogeny. *Biol. Rev.* **41**, 587–640.

Gould, S. J. 1967. Is uniformitarianism useful? *J. Geol. Educ.* **15**, 149–50.

Graf, W. L. 1977. The rate law in fluvial geomorphology. *Am. J. Sci.* **277**, 178–91.

Graf, W. L. 1979. Catastrophe theory as a model for change in fluvial systems. In *Adjustments of the fluvial system*, D. D. Rhodes & E. J. Williams (eds), 13–32. Binghamton Symp. Geomorph., Int. Ser., no. 10. London: Allen & Unwin.

Graf, W. L. 1982. Spatial variation of fluvial processes in semi-arid lands. In *Space and time in geomorphology*, C. E. Thorn (ed.), 193–217. Binghamton Symp. Geomorph., Int. Ser., no. 12. London: Allen & Unwin.

Gregory, K. J. & D. E. Walling 1973. *Drainage basin form and process: a geomorphological approach*. New York: Wiley.

Gregory, S. 1978. *Statistical methods and the geographer*, 4th edn. London: Longman.

Gretener, P. E. 1967. Significance of the rare event in geology. *Am. Assoc. Petrolm Geol. Bull.* **51**, 2197–206.

Griffiths, J. C. 1961. Measurement of the properties of sediment. *J. Geol.* **69**, 487–98.

Hack, J. T. 1960. Interpretation of erosional topography in humid temperate regions. *Am. J. Sci.,* **258–A**, 80–97 (Bradley Vol.).

Hack, J. T. 1975. Dynamic equilibrium and landscape evolution. In *Theories of landform development*, W. N. Melhorn & R. C. Flemal (eds), 87–102. Boston: Allen & Unwin.

Hack, J. T. 1980. *Rock control and tectonism – their importance in shaping the Appalachian Highlands*. U.S. Geol. Surv. Prof. Pap., no. 1126B.

Hack, J. T. 1982. *Physiographic divisions and differential uplift in the Piedmont and Blue Ridge*. U.S. Geol. Surv. Prof. Pap., no. 1265.

Haggett, P. 1965. Scale components in geographical problems. In *Frontiers in geographical teaching*, R. J. Chorley & P. Haggett (eds), 164–85. London: Methuen.

Haggett, P., R. J. Chorley, & D. R. Stoddart 1965. Scale standards in geographical research; a new measure of areal magnitude. *Nature* **205**, 844–7.

Haines-Young, R. H. & J. R. Petch 1980. The challenge of critical rationalism for methodology in physical geography. *Prog. Phys. Geog.* **4**, 63–78.

Haines-Young, R. H. & J. R. Petch 1983. Multiple working hypotheses: equifinality and the study of landforms. *Trans. Inst. Br. Geog.* N.S. **8**, 458–66.

Haines-Young, R. H. & J. R. Petch 1986. *Physical geography: its nature and methods*. London: Harper & Row.

Hall, A. D. & R. E. Fagen 1956. Definition of system. *Gen. Syst. Yearbook* **1**, 18–28.

Hallam, A. 1983. *Great geological controversies*. Oxford: Oxford University Press.

Hansel, A. K. 1980. *Sinkhole form as an indicator of process in karst landscape evolution*. Unpublished Ph.D. Dissertation, University of Illinois at Urbana-Champaign.

Harris, C. 1974. Wind speed and sand movement in a coastal dune environment. *Area* **6**, 243–9.

Harris, S. A. & C. R. Twidale 1968. Cycles, geomorphic. In *The encyclopedia of geomorphology*, R. W. Fairbridge (ed.), 237–40. New York: Reinhold.

Harvey, D. 1969. *Explanation in geography*. London: Edward Arnold.

Heylmun, E. B. 1971. Should we teach uniformitarianism? *J. Geol. Educ.* **19**, 35–7.

Higgins, C. G. 1975. Theories of landscape development: a perspective. In *Theories of landform development*, W. N. Melhorn & R. C. Flemal (eds), 1–28. Boston: Allen & Unwin.

High, C. & F. K. Hanna 1970. *A method for the direct measurement of erosion on rock surfaces*. Br. Geomorph. Res. Geoup Tech. Bull., no. 5.

Horton, R. E. 1932. Drainage basin characteristics. In *Am. Geophys. Union, Trans. 12th Annu. Meeting*, 350–61.

Horton, R. E. 1945. Erosional development of streams and their drainage basins; hydrophysical approach to quantitative morphology. *Bull. Geol. Soc. Am.* **56**, 275–370.

Howard, A. D. 1965. Geomorphological systems – equilibrium and dynamics. *Am. J. Sci.* **263**, 302–12.

Hsü, K. J. 1986. Commentary: Darwin's three mistakes. *Geology* **14**, 532–4.

Hubbert, M. K. 1967. Critique of the principle of uniformity. In *Uniformity and simplicity*, C. C. Albritton, Jr. (ed.), 3–34. Geol. Soc. Am., Spec. Pap., no. 89.

Huggett, R. J. 1985. *Earth surface systems*. Berlin: Springer.

Hurst, H. E. 1951. Long-term storage capacity of reservoirs. *Trans. Am. Soc. Civil Eng.* **116**, 770–808.

Huschke, R. E. (ed.) 1959. *Glossary of meteorology*. Boston: American Meteorological Society.

Hutton, J. 1788. Theory of the earth. *Trans. R. Soc. Edin.* **1**, 209–304.

Isaacson, E. de St Q. & M. de St Q. Isaacson 1975. *Dimensional methods in engineering and physics*. New York: Wiley.

Jenny, H. 1941. *Factors of soil formation*. New York: McGraw-Hill.

Johnson, A. M. 1970. *Physical processes in geology*. San Francisco: Freeman, Cooper.

Johnson, D. 1933. Role of analysis in scientific investigation. *Bull. Geol. Soc. Am.* **44**, 461–93.

Johnson, D. W. 1931. *Stream sculpture on the Atlantic slope, a study in the evolution of Appalachian rivers*. New York: Columbia University Press.

Judson, S. 1965. Quaternary processes in the Atlantic coastal plain and Appalachian highlands. In *The Quaternary of the United States*, H. E. Wright Jr. & D. G. Frey (eds), 133–6. Princeton, NJ: Princeton University Press.

Judson, S. 1975. Evolution of Appalachian topography. In *Theories of landform development*, W. N. Melhorn & R. C. Flemal (eds), 29–44. Boston: Allen & Unwin.

Kennedy, B. A. 1965. *An analysis of the factors influencing slope development on the Charmouthien Limestone of the Plateau de Bassigny, Haute-Marne, France*. B.A. Thesis, Department of Geography, Cambridge University.

Kennedy, B. A. 1980. A naughty world. *Trans. Inst. Br. Geog.* N.S. **4**, 550–8.

King, C. A. M. 1966. *Techniques in geomorphology*. London: Edward Arnold.

King, C. A. M. 1970. Feedback relationships in geomorphology. *Geog. Ann.* **52A**, 147–59.

King, C. A. M. 1980. *Physical geography*. Totowa, NJ: Barnes & Noble.

King, L. C. 1957. The uniformitarian nature of hillslopes. *Edin. Geol. Soc. Trans.* **17**, 81–104.

King, L. C. 1963a. Canons of landscape evolution. *Bull. Geol. Soc. Am.* **64**, 721–51.

King, L. C. 1963b. *South African scenery*, 3rd edn. Edinburgh: Oliver & Boyd. (First edition published 1942.)

King, L. C. 1967. *The morphology of the Earth*. New York: Hafner.

King, L. C. 1983. *Wandering continents and spreading sea floors on an expanding Earth*. Chichester: Wiley.

King, P. B. & S. A. Schumm 1980. *The physical geography (geomorphology) of William Morris Davis*. Norwich: Geo Books.

Kirkby, M. J. 1971. Hillslope process–response models based on the continuity equation. In *Slopes: form and process*, D. Brunsden (ed.), 15–30. Inst. Br. Geog. Spec. Pub., no. 3.

Kirkby, M. J. 1976a. Deterministic continuous slope models. *Z. Geomorph. N.F. Suppl.* **25**, 1–19.

Kirkby, M. J. 1976b. Hydrological slope models: the influence of climate. In *Geomorphology and climate*, E. Derbyshire (ed.), 247–67. London: Wiley.

Kirkby, M. J. 1977. Soil development models as a component of slope models. *Earth Surf. Processes* **2**, 203–30.

Kirkby, M. J. 1985. A model for the evolution of regolith-mantled slopes. In *Models in geomorphology*, M. J. Woldenberg (ed.), 213–37. Binghamton Symp. Geomorph., Int. Ser., no. 14, Boston: Allen & Unwin.

Kirkby, M. J. 1986. A two-dimensional simulation model for slope and stream evolution. In *Hillslope processes*, A. D. Abrahams (ed.), 205–22. Binghamton Symp. Geomorph., Int. Ser., no. 16. Boston: Allen & Unwin.

Klimaszewski, M. 1961. The problems of the geomorphological and hydrographic map on the example of the upper Silesian industrial district. *Prob. Appl. Geog., Geog. Stud.* (Poland) **25**, 73–81.

Knighton, D. 1984. *Fluvial forms and processes*. London: Edward Arnold.

Knox, J. C. 1972. Valley alluviation in southwestern Wisconsin. *Ann. Assoc. Am. Geog.* **62**, 401–10.

Knox, J. C. 1975. Concept of the graded stream. In *Theories of landform development*, W. N. Melhorn & R. C. Flemal (eds), 169–98. Boston: Allen & Unwin.

Krumbein, W. C. & F. A. Graybill 1965. *An introduction to statistical models in geology*. New York: Wiley.

Kuhn, T. S. 1970. *The structure of scientific revolutions*. Chicago: University of Chicago Press.

Langbein, W. B. & L. B. Leopold 1966. *River meanders – theory of minimum variance*. U.S. Geol. Surv. Prof. Pap., no. 422–H.

Le Ba Hong & T. R. H. Davies 1979. A study of stream braiding: summary. *Bull. Geol. Soc. Am.* **90**, 1094–5.

Lee, D. R. & G. T. Sallee 1970. A method of measuring shape. *Geog. Rev.* **60**, 555–63.

Lehmann, O. 1933. Morphologische theorie der Verwitterung von steinschlag wänden. *Vierteljahrsschr. Naturf. Ges. Zurich* **87**, 83–126.

Leopold, L. B., R. A. Bagnold, M. G. Wolman & L. M. Brush 1960. *Flow resistance in sinuous or irregular channels*. U.S. Geol. Surv. Prof. Pap., no. 282D.

Leopold, L. B. & T. Dunne 1971. *Field method for hillslope description*. Br. Geomorph. Res. Group. Tech. Bull., no. 7.

Leopold, L. B. & W. B. Langbein 1962. *The concept of entropy in landscape evolution*. U.S. Geol. Surv. Prof. Pap., no. 500–A.

Leopold, L. B. & T. Maddock Jr. 1953. *The hydraulic geometry of stream channels and some physiographic implications*. U.S. Geol. Surv. Prof. Pap., no. 252.

Leopold, L. B. & M. G. Wolman 1957. *River patterns: braided, meandering and straight*. U.S. Geol. Surv. Prof. Pap., no. 282–B.

Leopold, L. B., M. G. Wolman & J. P. Miller 1964. *Fluvial processes in geomorphology*. San Francisco: W. H. Freeman.

Lewin, J. 1980. Available and appropriate timescales in geomorphology. In *Timescales in geomorphology*, R. A. Cullingford, D. A. Davidson & J. Lewin (eds), 3–10. Chichester: Wiley.

Lewis, W. V. 1936. Nivation, river grading and shoreline development in southeast Iceland. *Geog. J.* **88**, 431–7.

Lewis, W. V. 1939. Snow-patch erosion in Iceland. *Geog. J.* **94**, 153–61.

Longwell, C. R. & R. F. Flint 1962. *Introduction to physical geology*, 2nd edn. New York: Wiley.

Lowman, S. W., et al. 1972. *Definitions of selected ground-water terms – revisions and conceptual refinements*. Geol. Surv. Water-Supply Paper, no. 1988.

Łoziński, W. 1909. Über die mechanische Verwitterung der Sandsteine im gemässigten Klima. *Acad. Sci. Cracovie Bull. Int., Cl. Sci. Math. Nat.* **1**, 1–25.

Lyell, C. 1830. *Principles of geology*, Vol. 1, 1st edn. Reprinted 1969. New York: Johnson Reprint Corp.

Lyubimov, B. P. 1967. On the mechanism of nival processes. *Podzemn. Led.* **III** (3), 158–75.

Mackin, J. H. 1948. Concept of the graded river. *Bull. Geol. Soc. Am.* **59**, 463–512.

Maddock, Jr., T. 1970. Indeterminate hydraulics of alluvial channels. *J. Hydraul. Div. Am. Soc. Civ. Engs.* **96**, HY11, 2309–23.

Madole, R. F. 1982. *Possible origins of till-like deposits near the summit of the Front Range in north-central Colorado*. U.S. Geol. Surv. Prof. Pap., no. 1243.

Mandelbrot, B. B. 1982. *The fractal geometry of nature*. San Francisco: W. H. Freeman.

Marchand, D. E. 1978. Quaternary deposits and Quaternary history. In *Quaternary deposits and soils of the central Susquehanna Valley of Pennsylvania*, 1–19. Guidebook 41st Ann. Reunion Northeast Friends of the Pleistocene.

Mardia, K. V. 1972. *Statistics of directional data*. London: Academic Press.

Margenau, H. 1961. *Open vistas*. New Haven, CN: Yale University Press.

Mark, D. M. 1974. On the interpretation of till fabrics. *Geology* **2**, 101–4.

Mark, D. M. 1975. Geomorphometric parameters: a review and evaluation. *Geog. Ann.* **57A**, 165–77.

Mark, D. M. 1983. Relations between field-surveyed channel networks and map-based geomorphometric measures, Inez, Kentucky. *Ann. Assoc. Am. Geog.* **73**, 358–72.

Mather, P. M. 1979. Theory and quantitative methods in geomorphology. *Prog. Phys. Geog.* **3**, 471–87.

Matthes, F. E. 1900. Glacial sculpture of the Bighorn Mountains, Wyoming. In *U.S. Geological Survey 21st Annual Report, 1899–1900*, 167–90.

McArthur, D. S. & R. Ehrlich 1977. An efficiency evaluation of four drainage basin shape ratios. *Prof. Geog.* **29**, 290–5.

McConnell, H. & J. M. Horn 1972. Probabilities of surface karst. In *Spatial analysis in geomorphology*, R. J. Chorley (ed.), 111–33. New York: Harper & Row.

McEwen, R. B., H. W. Calkins & B. S. Ramey 1983. *USGS digital cartographic data standards. Overview and USGS activities*. U.S. Geol. Surv. Circ., no. 895–A.

McKee, E. D. 1979. Sedimentary structures in dunes (with two sections on the Lagoa dune field, Brazil, by J. J. Bigarella). In *A study of global sand seas*, E. D. McKee (ed.), 83–134. U.S. Geol. Surv. Prof. Pap., no. 1052.

Melhorn, W. N. & D. E. Edgar 1975. The case for episodic, continental-scale erosion surfaces. In *Theories of landform development*, W. N. Melhorn & R. C. Flemal (eds), 243–76. Boston: Allen & Unwin.

Melton, M. A. 1958a. Correlation structure of morphometric properties of drainage systems and their controlling agents. *J. Geol.* **66**, 422–60.

Melton, M. A. 1958b. Geometric properties of mature drainage systems and their representation in E4 phase space. *J. Geol.* **66**, 35–54.

Miller, R. L. & J. S. Kahn 1962. *Statistical analysis in the geological sciences*. New York: Wiley.

Miller, T. K. & L. J. Onesti 1979. The relationship between channel shape and sediment characteristics in the channel perimeter. *Bull. Geol. Soc. Am.* **90**, 301–4.

Miller, V. C. 1953. *A quantitative geomorphic study of drainage basin characteristics in the Clinch Mountain area, Virginia and Tennessee*. Office Naval Res., Tech. Rep., no. 3. New York: Columbia University Press.

Morisawa, M. 1958. Measurement of drainage-basin outline form. *J. Geol.* **66**, 587–91.

Morisawa, M. 1985. *Rivers*. Geomorph. Texts, no. 7. London: Longman.

Morisawa, M. & J. T. Hack (eds) 1984. *Tectonic geomorphology*. Binghamton Symp. Geomorph. Int. Ser., no. 15. Boston: Allen & Unwin.

Mosley, M. P. & G. L. Zimpfer 1978. Hardware models in geomorphology. *Prog. Phys. Geog.* **2**, 438–61.

Murdoch, W. M. 1966. Community structure, population control, and competition – a critique. *Am. Nat.* **100**, 219–26.

Murphy, N. F. 1949. *Dimensional analysis*. Bull. Virginia Polytech. Inst. vol. 42, no. 6.

Nash, D. B. 1984. Morphologic dating of fluvial terrace scarps and fault scarps near West Yellowstone, Montana. *Bull. Geol. Soc. Am.* **95**, 1413–24.

Nystuen, J. D. 1963. Identification of some fundamental spatial concepts. *Pap. Michigan Acad. Sci., Arts, Lett.* **48**, 373–84.

Odum, H. T. 1972. An energy circuit language for ecological and social systems: its physical basis. In *Systems analysis and simulation in ecology*, Vol. 2, B. C. Patten (ed.), 139–211. New York: Academic Press.

Ogden, C. K. & I. A. Richards 1949. *Meaning of meaning*, 2nd edn. London: Routledge & Kegan Paul.

Ollier, C. D. 1977. Terrain classification: methods, applications and principles. In *Applied geomorphology*, J. R. Hails (ed.), 277–316. Amsterdam: Elsevier.

Ollier, C. D. 1984. Morphotectonics of continental margins with great escarpments. In *Tectonic geomorphology*, M. Morisawa & J. T. Hack (eds), 3–25. Binghamton Symp. Geomorph., Int. Ser., no. 15. Boston: Allen & Unwin.

Ongley, E. D. 1970. Drainage basin axial and shape parameters from moment measures. *Can. Geog.* **14**, 38–44.

Paine, A. D. M. 1985. "Ergodic" reasoning in geomorphology: time for a review of the term? *Prog. Phys. Geog.* **9**, 1–15.

Parker, R. B. 1985. Buffers, energy storage, and the mode and tempo of geologic events. *Geology* **13**, 440–2.

Parker, R. S. 1977. *Experimental study of drainage basin evolution and its hydrologic implications*. Colorado State Univ. Hydrol. Pap., no. 90.

Parry, W. 1981. *Topics in ergodic theory*. Cambridge: Cambridge University Press.

Peltier, L. C. 1950. The geographic cycle in periglacial regions as it is related to climatic geomorphology. *Ann. Assoc. Am. Geog.* **40**, 214–36.

Peltier, L. C. 1975. The concept of climatic geomorphology. In *Theories of landform development*, W. N. Melhorn & R. C. Flemal (eds), 129–43. Boston: Allen & Unwin.

Penck, W. 1924. *Die morphologische Analyse: Ein Kapital der physikalischen Geologie*. Geog. Abh., 2 Reihe, Heft 2. Stuttgart: Engelhorn.

Penck, W. 1925. *Die piedmontflächen des südlichen Schwarzwaldes*, 81–108. Z. Ges. Erdk. Berlin.

Petch, J. R. & R. H. Young 1978. *The challenge of deductive science for geographical investigation*. Univ. Salford, Disc. Pap. Geog., no. 7.

Pike, R. J. & S. E. Wilson 1971. Elevation–relief ratio, hypsometric integral, and geomorphic area–altitude analysis. *Bull. Geol. Soc. Am.* **82**, 1079–83.

Pitty, A. F. 1982. *The nature of geomorphology*. London: Methuen.

Popper, K. R. 1972. *The logic of scientific discovery*, 6th revised impression. London: Hutchinson.

Press, F. & R. Siever 1982. *Earth*, 3rd edn. San Francisco: W. H. Freeman.

Prigogine, I. 1980. *From being to becoming: time and complexity in the physical sciences*. San Francisco: W. H. Freeman.

Pyne, S. J. 1980. *Grove Karl Gilbert – A great engine of research*. Austin, TX: University of Texas Press.

Rapp, A. 1960. Recent development of mountain slopes in Kärkevagge and surroundings, north Scandinavia. *Geog. Ann.* **42**, 65–200.

Ray, R. J., W. B. Krantz, T. N. Caine & R. D. Gunn 1983. A model for sorted patterned-ground regularity. *J. Glaciol.* **29**, 317–37.

Reynaud, A. 1971. *Épistémologie de la géomorphologie*. Paris: Masson.

Reynolds, W. C. 1974. *Energy: from nature to man*. New York: McGraw-Hill.

Richards, K. 1982. *Rivers: form and process in alluvial rivers*. London: Methuen.

Riley, N. A. 1941. Projection sphericity. *J. Sed. Petrol.* **11**, 94–7.

Ruhe, R. V. 1969. *Quaternary landscapes in Iowa*. Ames, IA: Iowa State University Press.

Ryder, J. M. 1971. Some aspects of the morphometry of paraglacial alluvial fans in south-central British Columbia. *Can. J. Earth Sci.* **8**, 1252–64.

Sapir, E. 1921. *Language: an introduction to the study of speech*. New York: Harcourt, Brace.

Sauchyn, D. J. & J. S. Gardiner 1983. Morphometry of open rock basins, Kananaskis area, Canadian Rocky Mountains. *Can. J. Earth Sci.* **20**, 409–19.

Saull, V. A. 1986. Wanted: alternatives to plate tectonics. *Geology* **14**, 536.

Savigear, R. A. G. 1952. Some observations on slope development in South Wales. *Trans. Inst. Br. Geog.* **18**, 31–52.

Savigear, R. A. G. 1965. A technique of morphological mapping. *Ann. Assoc. Am. Geog.* **43**, 514–38.

Scheidegger, A. E. 1961. Mathematical models of slope development. *Bull. Geol. Soc. Am.* **72**, 37–50.

Scheidegger, A. E. 1970. *Theoretical geomorphology*, 2nd edn. Berlin: Springer.

Schumm, S. A. 1956. Evolution of drainage systems and slopes in badlands at Perth Amboy, New Jersey. *Bull. Geol. Soc. Am.* **67**, 597–646.

Schumm, S. A. 1963a. Sinuosity of alluvial rivers on the Great Plains. *Bull. Geol. Soc. Am.* **74**, 1089–100.

Schumm, S. A. 1963b. *The disparity between present rates of denudation and orogeny*. U.S. Geol. Surv. Prof. Pap., no. 454-H.

Schumm, S. A. 1979. Geomorphic thresholds: the concept and its applications. *Trans. Inst. Br. Geog. N.S.* **4**, 485–515.

Schumm, S. A. 1985. Explanation and extrapolation in geomorphology: seven reasons for geologic uncertainty. *Trans. Jap. Geomorph. Union* **6**, 1–18.

Schumm, S. A. & R. W. Lichty 1965. Time, space, and causality in geomorphology. *Am. J. Sci.* **263**, 110–19.

Selby, M. J. 1976. Slope erosion due to extreme rainfall: a case study from New Zealand. *Geog. Ann.* **58A**, 131–8.

Selby, M. J. 1982. *Hillslope materials and processes*. Oxford: Oxford University Press.

Shea, J. H. 1982. Twelve fallacies of uniformitarianism. *Geology* **10**, 455–60.

Shea, J. H. 1983. Creationism, uniformitarianism, geology, and science. *J. Geol. Educ.* **31**, 105–10.

Shelton, J. S. 1966. *Geology illustrated*. San Francisco: W. H. Freeman.

Shreve, R. L. 1966. Statistical law of stream numbers. *J. Geol.* **74**, 17–37.

Shreve, R. L. 1967. Infinite topologically random channel networks. *J. Geol.* **75**, 178–86.

Shreve, R. L. 1969. Stream lengths and basin areas in topologically random channel networks. *J. Geol.* **77**, 397–414.

Siegel, S. 1956. *Nonparametric statistics*. New York: McGraw-Hill.

Simons, M. 1962. The morphological analysis of landforms: a new review of the work of Walter Penck (1888–1923). *Trans. Inst. Br. Geog.* **31**, 1–14.

Simpson, G. G. 1963. Historical science. In *The fabric of geology*, C. C. Albritton Jr. (ed.), 24–48. Stanford, CA: Freeman, Cooper.

Slaymaker, O., T. Dunne & A. Rapp 1980. Geomorphic experiments on hillslopes. *Z. Geomorph, N.F. Suppl.* **35**, v–vii.

Smalley, I. J. & C. Vita-Finzi 1969. The concept of "system" in the earth sciences, particularly geomorphology. *Bull. Geol. Soc. Am.* **80**, 1591–4.

Smyth, D. S. & P. B. Checkland 1976. Using a systems approach: the structure of root definitions. *J. Appl. Syst. Anal.* **5**, 75–83.

Stanford, S. D. 1983. Fabric and depositional structures in drumlins near Waukesha, Wisconsin. *Geosci. Wisc.* **7**, 98–111.

Starkel, L. 1976. The role of extreme (catastrophic) meteorological events in contemporary evolution of slopes. In *Geomorphology and climate*, E. Derbyshire (ed.), 203–46. London: Wiley.

Stoddart, D. R. 1966. Darwin's impact on geography. *Ann. Assoc. Am. Geog.* **56**, 683–98.

Stoddart, D. R. 1969. Climatic geomorphology: review and re-assessment. *Prog. Geog.* **1**, 159–222.

Strahler, A. N. 1950. Equilibrium theory of erosional slopes approached by frequency distribution analysis. *Am. J. Sci.* **248**, 673–96 and 800–14.

Strahler, A. N. 1952a. Dynamic basis of geomorphology. *Bull. Geol. Soc. Am.* **63**, 923–38.

Strahler, A. N. 1952b. Hypsometric (area–altitude) analysis of erosional topography. *Bull. Geol. Soc. Am.* **63**, 1117–42.

Strahler, A. N. 1957. Quantitative analysis of watershed geomorphology. *Trans. Am. Geophys. Union* **38**, 913–20.

Strahler, A. N. 1958. Dimensional analysis applied to fluvially eroded landforms. *Bull. Geol. Soc. Am.* **69**, 279–300.

Strahler, A. N. 1969. *Physical geography*, 3rd edn. New York: Wiley.

Strahler, A. N. 1971. *The Earth sciences*, 2nd edn. New York: Harper & Row.

Strahler, A. N. 1980. Systems theory in physical geography. *Phys. Geog.* **1**, 1–27.

Strahler, A. N. & A. H. Strahler 1973. *Environmental geoscience: interaction between natural systems and man*. Santa Barbara, CA: Hamilton.

Suess, E. 1883–1908. *Das Antlitz der Erde*, 3 Vols. Wien: Tempsky.

Summerfield, M. A. 1984. Plate tectonics and landscape development on the African continent. In *Tectonic geomorphology*, M. Morisawa & J. T. Hack (eds), 27–51. Binghamton Symp. Geomorph., Int. Ser., no. 15. Boston: Allen & Unwin.

Teichert, C. 1958. Concepts of facies. *Bull. Am. Assoc. Petrolm Geol.* **42**, 2718–44.

Terjung, W. H. 1976. Climatology for geographers. *Ann. Assoc. Am. Geog.* **66**, 199–222.

Thom, R. 1975. *Structural stability and morphogenesis.* New York: Benjamin.

Thomas, E. 1965. *A structure of geography: a proto-unit for secondary schools.* High School Geog. Proj., Assoc. Am. Geog. Boulder, CO: University of Colorado.

Thompson, D. W. 1961. *On growth and form*, abridged edn., J. T. Bonner (ed.). Cambridge: Cambridge University Press.

Thorn, C. E. 1976. Quantitative evaluation of nivation in the Colorado Front Range. *Bull. Geol. Soc. Am.* **87**, 1169–78.

Thorn, C. E. 1982. Gopher disturbance: its variability by Braun-Blanquet vegetation units in the Niwot Ridge alpine tundra zone, Colorado Front Range, U.S.A. *Arctic Alpine Res.* **14**, 45–51.

Thorn, C. E. 1988. Nivation: a geomorphic chimera. In *Advances in Periglacial geomorphology*, M. J. Clark (ed.), 3–31. Chichester: Wiley.

Thornbury, W. D. 1965. *Regional geomorphology of the United States.* New York: Wiley.

Thornbury, W. D. 1969. *Principles of geomorphology*, 2nd edn. New York: Wiley.

Thornes, J. 1979. Fluvial processes. In *Processes in geomorphology*, C. Embleton & J. Thornes (eds), 213–71. New York: Wiley.

Thornes, J. B. 1983. Evolutionary geomorphology. *Geography* **68**, 225–35.

Thornes, J. B. & D. Brunsden 1977. *Geomorphology and time.* New York: Wiley.

Thornes, J. B. & R. I. Ferguson 1981. Geomorphology. In *Quantitative geography: a British view*, N. Wrigley & R. J. Bennett (eds), 284–93. London: Routledge & Kegan Paul.

Till, R. 1974. *Statistical methods for the Earth scientist.* London: Macmillan.

Tinkler, K. J. 1985. *A short history of geomorphology.* London: Croom Helm.

Tricart, J. & A. Cailleux 1972. *Introduction to climatic geomorphology.* London: Longman.

Trimble, S. W. 1975. Denudation studies: can we assume stream steady state? *Science* **188**, 1207–8.

Trimble, S. W. 1977. The fallacy of stream equilibrium in contemporary denudation studies. *Am. J. Sci.* **277**, 876–87.

Tuan, Y.-F. 1958. The misleading antithesis of Penckian and Davisian concepts of slope retreat in waning development. *Proc. Indiana Acad. Sci.* **67**, 212–14.

Twidale, C. R. 1976. On the survival of paleoforms. *Am. J. Sci.* **276**, 77–95.

Twidale, C. R. 1977. Fragile foundations: some methodological problems in geomorphological research. *Rev. Géomorph. Dyn.* **26**, 84–95.

Twidale, C. R. 1983. Pediments, peneplains and ultiplains. *Rev. Géomorph. Dyn.* **32**, 1–35.

Twidale, C. R. 1987. Etch and intracutaneous landforms and their implications. *Aust. J. Earth Sci.* **34**, 367–86.

United States Army, Corps of Engineers 1949. *Unit hydrograph compilations.* Washington District, project CW 153.

United States Department of Agriculture 1975. *Soil taxonomy – a basic system of soil classification for making and interpreting soil surveys.* U.S. Dept. Agric. Soil Conserv. Serv. Agric. Handb., no. 436.

Verstappen, H. T. 1977. *Remote sensing in geomorphology.* Amsterdam: Elsevier.

Vita-Finzi, C. 1973. *Recent Earth history.* New York: Wiley.

von Bertalanffy, L. 1950. An outline of general system theory. *Br. J. Phil. Sci.* **1**, 134–65.

von Bertalanffy, L. 1956. General system theory. *Gen. Syst.* **1**, 1–10.

von Bertalanffy, L. 1962. General system theory – a critical review. *Gen. Syst.* **7**, 1–20.

von Engeln, O. D. 1942. *Geomorphology – systematic and regional.* New York: Macmillan.

Washburn, A. L. 1980. *Geocryology: a survey of periglacial processes and environments.* New York: Wiley.

Washburn, A. L., J. E. Sanders & R. F. Flint 1963. A convenient nomenclature for poorly sorted sediments. *J. Sed. Petrol.* **33**, 478–80.

Watson, R. A. 1966. Is geology different? A critical discussion of "The fabric of geology". *Philos. Sci.* **33**, 172–85.

Watson, R. A. 1969. Explanation and prediction in geology. *J. Geol.* **77**, 488–94.

Webster's dictionary of the English language, unabridged, encyclopedic edition 1977. Springfield, Mass.: G. & C. Merriam.

Weyman, D. R. 1974. Runoff process, contributing area and streamflow in a small upland catchment. In *Fluvial processes in instrumented watersheds*, K. J. Gregory & D. E. Walling (eds), 33–43. London: Institute of British Geographers.

Whalley, W. B. 1972. The description and measurement of sedimentary particles and the concept of form. *J. Sed. Petrol.* **42**, 961–5.

Whalley, W. B. 1976. *Properties of materials and geomorphological explanation*. London: Oxford University Press.

Whittecar, G. R. & D. M. Mickelson 1979. Composition, internal structures, and an hypothesis of formation of drumlins, Waukesha County, Wisconsin, U.S.A. *J. Glaciol.* **22**, 357–71.

Whorf, B. F. 1939. The relation of habitual thought and behavior to language. In *Language, culture and personality*, L. Spier (ed.), 75–93. Menasha, WI: Sapir Memorial Publication Fund.

Whorf, B. F. 1940a. Linguistics as an exact science. *Technol. Rev.* **43**, 61–3 and 80–3.

Whorf, B. F. 1940b. Science and linguistics. *Technol. Rev.* **42**, 229–31 and 247–8.

Whorf, B. F. 1941. Language and logic. *Technol. Rev.* **43**, 250–2, 266, 268 and 272.

Williams, P. J. 1982. *The surface of the Earth: an introduction to geotechnical science*. London: Longman.

Winkelmolen, A. M. 1982. Critical remarks on grain parameters, with special emphasis on shape. *Sedimentology* **29**, 255–65.

Woldenberg, M. J. 1972. The average hexagon in spatial hierarchies. In *Spatial analysis in geomorphology*, R. J. Chorley (ed.), 323–52. New York: Harper & Row.

Wolman, M. G. & R. Gerson 1978. Relative scales of time and effectiveness of climate in watershed geomorphology. *Earth Surf. Processes* **3**, 189–208.

Wolman, M. G. & J. P. Miller 1960. Magnitude and frequency of forces in geomorphic processes. *J. Geol.* **68**, 54–74.

Womack, W. R. & S. A. Schumm 1977. Terraces of Douglas Creek, northwestern Colorado: an example of episodic erosion. *Geology* **5**, 72–6.

Wood, A. 1942. The development of hillside slopes. *Proc. Geol. Assoc.* **53**, 128–40.

Wood, W. F. & J. B. Snell 1960. *A quantitative system for classifying landforms*. U.S. Dept. Army, Natick Lab., Tech. Rep., no. EP-124.

Wooldridge, S. W. & D. L. Linton 1955. *Structure, surface and drainage in south-east England*, 2nd edn. London: Philip.

Yang, C. T. 1971. On river meanders. *J. Hydrol.* **13**, 231–53.

Yang, C. T. & C. C. S. Song 1979. Dynamic adjustments of alluvial channels. In *Adjustments of the fluvial system*, D. D. Rhodes & G. P. Williams (eds), 55–67. Dubuque, IA: Kendall/Hunt.

Yatsu, E. 1966. *Rock control and geomorphology*. Tokyo: Sozisha.

Yeates, M. 1974. *An introduction to quantitative analysis in human geography*. New York: McGraw-Hill.

Young, A. 1972. *Slopes*. Edinburgh: Oliver & Boyd.

Young, A. 1974. *Slope profile survey*. Br. Geomorph. Res. Group Tech. Bull., no. 11.

Young, R. H. & J. R. Petch 1978. *The methodological limitations of Kuhn's model in science*. Univ. Salford, Disc. Pap. Geog., no. 8.

Subject Index

Numbers that are in italics refer to figures.

Author Index

Numbers that are in italics refer to figures.